SpringerBriefs in Systems Biology

W0193257

For further volumes:
http://www.springer.com/series/10426

Michael Kinter • Caroline S. Kinter

Application of Selected Reaction Monitoring to Highly Multiplexed Targeted Quantitative Proteomics

A Replacement for Western Blot Analysis

 Springer

Michael Kinter
Oklahoma Medical Research Foundation
Oklahoma City, OK, USA

Caroline S. Kinter
Oklahoma Medical Research Foundation
Oklahoma City, OK, USA

ISSN 2193-4746 ISSN 2193-4754 (electronic)
ISBN 978-1-4614-8665-7 ISBN 978-1-4614-8666-4 (eBook)
DOI 10.1007/978-1-4614-8666-4
Springer New York Heidelberg Dordrecht London

Library of Congress Control Number: 2013947406

Printed on acid-free paper

Springer is part of Springer Science+Business Media (www.springer.com)

For Lauren and Courtney

Preface

The original experiment in the field of proteomics identified proteins contained in SDS-PAGE gel bands by digesting those proteins with the enzyme trypsin and sequencing the resulting peptides by collision-induced dissociation in a tandem mass spectrometer. Although the reasons for selecting any particular gel band for identification varied, a common reason was because a change had been seen in the abundance of that band under different experimental conditions. While these experiments were certainly quantitative proteomics, the quantitative information came from the gel through the pattern of protein staining that was seen. In 2000, we wrote one of the first books describing this identification experiment on a practical level.

The goal of this book is to describe a new type of quantitative proteomics experiment called selected reaction monitoring that has seen growing use over the past several years. The selected reaction monitoring experiment leverages several developments in the fields of genomics, information technology, and mass spectrometry. The result is a robust and powerful method to measure the abundance of a protein in complex samples, with high specificity and sensitivity.

Oklahoma City, OK, USA   Michael Kinter
Oklahoma City, OK, USA   Caroline S. Kinter

Acknowledgments

There are three groups of people that contributed to the ideas and work presented in this volume. The Proteomics group in the Department of Cell Biology in the Lerner Research Institute of the Cleveland Clinic carried out the experiments that laid the groundwork for the current methods in the course of an extensive series of identification experiments and a modest number of quantitative experiments—Belinda Willard, Suma Kaveti, Andrew Keightley, Heather Hargett, Cristian Ruse, James Conway, Lemin Zheng, Assem Ziady, and Dongmei Zhang. More recently, our group in the Free Radical Biology and Aging Research Program at the Oklahoma Medical Research Foundation has now taken the targeted quantitative proteomics work and made it a routine, highly productive tool for our research program—Jolyn Fernandes, Clair Crewe, Paul Rindler, Halee Patel, Jillian Lundie, Alex Weddle, Aaron McLain, Melinda West, and Luke Szweda. In both institutions we have been fortunate to collaborate with a large group of scientists with outstanding research programs and challenging analytical problems that were an exceptional driving force throughout the refinement of these approaches. In addition, we would like to acknowledge Michael MacCoss at the University of Washington School of Medicine. His lunchtime presentation sponsored by ThermoScientific in June 2008 at the Annual Meeting of the American Society for Mass Spectrometry conveyed both the excitement and opportunity of using selected reaction monitoring in this way.

Contents

List of Abbreviations

ALiPHAT	Augmented limits of detection for peptides with hydrophobic alkyl tags
ATP	Adenosine-5′-triphosphate
AQUA	Absolute quantification
BSA	Bovine serum albumin
CID	Collision induced dissociation
DTT	Dithiolthreitol
EDTA	Ethylenediaminetetraacetic acid
ELISA	Enzyme-linked immunosorbent assay
eV	Electron volts
GC	Gas chromatography
ICAT	Isotope coded affinity tag
LC	Liquid chromatography
MIKES	Mass analyzed ion kinetic energy spectroscopy
MOPS	3-(N-morpholino)propanesulfonic acid
MRM	Multiple reaction monitoring
m/z	Mass-to-charge ratio
PAGE	Polyacrylamide gel electrophoresis
PCR	Polymerase chain reaction
RIPA	Radioimmunoprecipitation assay buffer
RNA	Ribonucleic acid
SDS	Sodium dodecyl sulfate
SILAC	Stable isotope labeling in culture
SIM	Selected ion monitoring
SISCAPA	Stable isotope standards and capture by anti-peptide antibodies
SRM	Selected reaction monitoring
SSRC	Sequence specific retention calculator
Tris	Tris(hydroxymethyl)aminomethane
TPR	Total protein response
QconCAT	Quantification concatamers

Chapter 1
Introduction to Quantitative Proteomics

Abstract The use of mass spectrometry for quantitative analysis began with the very first published experiment in 1913. Therefore, current applications, including quantitative proteomics, are based on an extensive foundation of fundamental and applied investigations. This history also applies to selected reaction monitoring. The ability of tandem mass spectrometry to enhance the specificity of an analysis was apparent in the 1970s and the technique has a strong history of use for many classes of small molecules. As a result, the application of selected reaction monitoring to targeted quantitative proteomics can be seen as a straight-forward experiment based on sound principles, excellent equipment, and a distinguished history.

1.1 The Origins of Quantitative Proteomics

Mass spectrometry is a naturally quantitative tool because the strength of the signal is directly proportional to the amount of the analyte responsible for that signal. This trait has been recognized since the beginning of the field. In his interpretation of the first series of mass spectra published in the first mass spectrometry paper, Thomson observed that neon-22 is less abundant than neon-20 in a sample of neon gas because, 'The parabola due to the heavier gas is always much fainter than that due to the lighter, so that probably the heavier gas forms only a small percentage of the mixture [1].' The challenges of quantitative mass spectrometry are the same as other quantitative methods, how to maximize the signal obtained from the compound being measured and how to assure that the signal is due to that compound alone. As Thomson recognized, mass spectrometry was especially suited to these challenges due to a powerful combination of the identity information in 'm/e' ratios, the ability to directly analyze mixtures since 'the presence of other gases is a matter of comparatively little importance', and finally the 'sensitiveness of the positive ray method' [1]. From this beginning in 1913, mass spectrometry has become a dominant tool for the quantitative analysis of essentially any class of molecule, including proteins.

M. Kinter and C.S. Kinter, *Application of Selected Reaction Monitoring to Highly Multiplexed Targeted Quantitative Proteomics*, SpringerBriefs in Systems Biology, DOI 10.1007/978-1-4614-8666-4_1, © The Authors 2013

Quantitative determinations were also a part of the original proteomic experiments. In those experiments, however, protein abundance was tracked by other types of information. The most common experimental paradigm was to monitor protein abundance in a gel electrophoresis experiment. Differentially expressed proteins bands were seen based on the pattern of staining and cut directly from that gel for identification by tandem mass spectrometry [2]. Subsequent validation experiments most often use the identification information to initiate Western blot analyses to verify the original abundance observation. Biomedical investigators continue to use this experimental approach with good results.

In running these identification experiments, many laboratories recognized that when multiple proteins were identified in one sample, the relative abundance of the proteins could be judged based on the number of collision induced dissociation (CID) spectra that matched each sequence. These observations became the basis of label-free quantitative proteomic experiments. Label-free methods were rapidly accepted because they do not require any additional manipulations of the sample, generating the abundance information through statistical analyses of the identification results [3]. More recent experiments commonly add signal strength information to augment the scan counting results [4]. The effectiveness of these experiments has been enhanced over the years by advances in mass spectrometry instrumentation and operation. Increased sensitivity, faster scan speeds, and routine high m/z resolution all work together to allow deeper interrogation of complex samples. The additional data obtained by these instruments gives a more complete picture of the sample proteome and improves the quantitative data with better counting statistics.

Other quantitative proteomic approaches utilize a key attribute of mass spectrometry — the ability to independently measure different isotopologues of a compound. One of the first isotope labeling methods was the isotope coded affinity tag (ICAT) experiment [5]. The ICAT experiment chemically labels the cysteines in different samples with isotopically labeled iodoacetamide derivatives. An innovative feature of the reagent was the inclusion of a biotin-moiety which allowed the complexity of the mixture to be reduced by affinity isolation of the labeled peptides. Other commonly used isotope labeling methods include labeling peptides with ^{18}O by digestion in ^{18}O-water [6] and peptide N-terminal labeling reagents [7]. Part of the utility of the chemical labeling methods is the availability of multiplexed systems composed of multiple isotopologues [7, 8]. The ultimate extension of the isotope labeling strategy for quantitative proteomics is the technique of stable isotope labeling in culture (SILAC) [9]. This labeling method limits the availability of the essential amino acids lysine and arginine to specific isotopologues to produce a set of cell lines in which all proteins have either the light or heavy versions of the amino acids incorporated into their proteins. Recently, a so-called SILAC mouse has been described for whole animal experiments [10]. In all of these quantitative methods, the quantitative comparison is based on mixing a sample generated using the light isotope version of the label with equal amounts of a sample generated using the heavy isotope version of the label prior to analysis by traditional data-dependent analyses. The quantitative information for each identified protein is then taken from the abundance ratio of the heavy isotope peptide to the light isotope peptide.

As a group, these different approaches to quantitative proteomics have been extraordinarily productive, giving unique insights into the function and dysfunction of a wide-ranging set of biological systems. These experiments have clearly demonstrated the utility of quantitative proteomics and the power of mass spectrometry experiments for generating this quantitative data. It is certain that these methods will continue to advance and remain important tools for detecting changes in protein expression. In this regard, the discussion of targeted quantitative proteomics by selected reaction monitoring is not based on real or perceived *weaknesses* in these other quantitative methods. Rather, the selected reaction monitoring experiment is best seen as new tool that complements those methods for an additional type of quantitative experiment.

1.2   The Role for Targeted Quantitation

Appreciating the role of targeted quantitative proteomics in broader field of quantitative proteomics begins with recognizing three features of the quantitative methods noted above. *First,* the use of data dependent analyses means that the mass spectrometer is deciding which ions to characterize. The power of data dependent analysis is the unbiased nature and the ability to discover new differentially regulated proteins. The ability to measure any one specific protein, however, is not assured. *Second,* the extensive data sets needed for the combination of deep interrogation and good counting statistics requires both elaborate acquisition strategies and powerful data analysis tools. This extensive characterization uses multidimensional fractionation combined with long LC-tandem mass spectrometry experiments [11]. The result is that many experiments require approximately 24 h of instrument time and produce 250,000 CID spectra in the analysis of a single sample. Multiple comparisons and replicate analyses then increase the time and number of CID spectra proportionally. *Third,* significant parts of the proteome are not accessed by these unbiased shotgun experiments. One estimate is that over 100,000 peptides can be detected by high resolution mass spectrometry during a capillary column LC separation of a cell lysate but less than 10,000 peptides were identified by the 22,000 CID spectra that were recorded [12].

The role of targeted quantitative proteomics is to fill the unmet need to use the power of mass spectrometry in quantitative analyses to measure the abundance of a specific subset of proteins. One measure of the success that targeted quantitative proteomics is enjoying in this role is the recognition by the journal Nature Methods as the 'Method of the Year 2012' [13]. An important part of the recognition of targeted proteomics was the step it takes beyond the traditional antibody-based experiments that are most commonly used for targeting proteins. These advantages include the faster development process of the selected reaction monitoring experiment relative to antibody development, superior specificity, and the ability to multiplex. These points are emphasized in the editorials that accompany the report [14, 15].

Of particular note is the ability of protein quantitation by selected reaction monitoring to replace Western blot analysis as a favored protein quantitation tool. Four critical problems with Western blot are direct strengths of selected reaction monitoring—the challenges of antibody production, the semi-quantitative nature of Western blot, its limited dynamic range, and the inability to analyzed large or even moderate numbers of samples. In each of these cases, selected reaction monitoring has the opposite traits. As discussed in Chap. 3, the development of a selected reaction monitoring assay is a deliberate process with new and expanding database resources. Once established, a good selected reaction monitoring method can be used forever. Selected reaction monitoring results are numerical and have a dynamic range of several orders of magnitude. Importantly, sample throughput and LC automation allow large numbers of samples to be analyzed. Therefore, the results of targeted quantitative proteomics include replicates and appropriate statistical testing, as opposed to the representative results approach often used with Western blot. The multiplex capabilities are also significant with methods measuring approximately 25 proteins in a single assay being routine and over a 100 proteins possible in many systems. It is also important to note that interlaboratory comparisons have been done and show both a transferability of methods and a comparability of results [16].

Finally, it is worthwhile to consider the costs of selected reaction monitoring experiments. While there is no doubt that advanced capillary column LC-triple quadrupole mass spectrometers are far more expensive than gel electrophoresis instruments, the mass spectrometer purchase price is not the complete story. The throughput of the selected reaction monitoring experiment, when considering the number of samples and the number of proteins assayed in each sample, is remarkably high. For example, in 2012 our laboratory analyzed over 1,000 samples for approximately 100 proteins per sample to give a total of 100,000 quantitative determinations. Using a cost of $500,000 to purchase, operate, and supply our instrument system for the year, the cost per protein measurement is just over $5 each. Importantly, the instrument was *not consumed* in this process and remains available to carry out another 100,000 determinations in 2013 (and beyond) with the on-going costs limited to instrument maintenance and assay supplies. By comparison, a typical Western blot antibody costs approximately $300 for an amount suitable for 100 blots. If 10 samples are included in every blot, then the antibody costs are $0.30 per sample per protein. This translates to $30 in antibody costs per sample for the 100 proteins we measure by selected reaction monitoring. Further, the antibodies are consumed in the process and must be repurchased the following year. While this type of costs analysis has many broad assumptions made for illustrative purposes, the simple point to be made is this—while mass spectrometers do have a high purchase price, effective utilization can give a proportionally high return and make the effective costs of individual experiments competitive with other methods.

1.3   A Brief History of Selected Reaction Monitoring

As discussed in greater detail in Chap. 2, the concept of using tandem mass spectrometry to give a higher level of specificity in an analysis was recognized in the first tandem mass spectrometry experiments. Since the initial development of mass analyzed ion kinetic energy spectroscopy (MIKES) in the 1970s, mass spectroscopists have used tandem mass spectrometry to characterize the structure of chemical compounds [17]. Collision induced dissociation added to this capability [18], as did the development of the triple quadrupole tandem mass spectrometer [19]. A part of these early applications were demonstrations of both qualitative and quantitative analyses in mixtures [20, 21]. Often these direct analyses were discussed in the context of replacing chromatography when analyzing complex mixtures, but the advantages of retaining the chromatography even in a very rapid fashion was also recognized [22, 23]. In many ways, this combination of rapid chromatography and tandem mass spectrometry for high throughput analyses is a key development that has led to the broad use of tandem mass spectrometry in quantitative analyses. Specifically, this speed of analysis has given LC-tandem mass spectrometry a place in a variety of analytical settings including pharmaceutical, environment, food testing, and clinical laboratories (to name a few). Key effects of these wide-ranging applications are that selected reaction monitoring is a common experiment with a broad base of users that understand its power and utility.

In the world of proteomics, selected reaction monitoring initially had a different connotation because of the common use of ion trap instruments in protein identification experiments. Ion traps have a slow duty cycle because of the trapping phase of the m/z scan. As a result, most uses of the selected reaction monitoring approach were still in the area of qualitative analyses. These applications were often described as hypothesis-testing mass spectrometry to reflect the fact that the instrument was used to probe for a specific structure in a mixture based on other forms of preliminary data [24, 25]. Our laboratory used this selected reaction monitoring approach for a number of qualitative analyses, generally with the goal of finding a phosphorylated peptide [26]. We eventually recognized that a similar approach could be used for quantitative analyses with an ion trap instrument despite the slow scan speeds [27, 28].

The idea that selected reaction monitoring could become a routine tool for protein quantitation began to accelerate in the early 2000s, with an initial report suggesting several advantages of tandem mass spectrometry relative to Western blot analysis [29]. Other papers introduced the combination of selected reaction monitoring and isotope dilution for the absolute quantitation of proteins, including an assay of the important clinical biomarker prostate specific antigen [30, 31]. These reports are also notable because they began to use the triple quadrupole mass spectrometer in the place of an ion trap system. The breadth of the analysis was soon increased by higher degrees of multiplexing [32]. A result of these successful analyses of low abundance proteins in human blood samples helped drive recognition of selected reaction monitoring as an ideal quantitative tool in the validation stages of biomarker discovery efforts [33, 34]. The final critical addition throughout these early applications of selected reaction monitoring was the development of the various software and informatics tools needed to make the design, execution, and data analysis of the experiment routine [35, 36].

1.4 Purpose of This Book

The purpose of this book is to describe a combination of fundamental and practical aspects of the selected reaction monitoring experiment in a way that will help a new laboratory begin using the method as quickly as possible. Because of the brief format of the book, we have not attempted an extensive literature review. Instead, we have focused on a small number of papers to support the relevant issues and give an entry into the literature. We have also limited the scope of the book to the basics concepts and steps needed to get the functional experiments started. As a consequence, there are a number of more advanced topics that remain uncovered.

The descriptions use an essentially personal approach that is based on our own laboratory's experience with the experiment. As with any advanced technique, some of the choices we make are different from what other laboratories might choose. We cannot say that all our methods are absolutely the best, but we can say that they have worked well over a nearly 5-year period of time in the analysis of a several thousand samples derived from a variety of sample types. There are also places where the description we present may seem more elaborate or tedious than necessary, especially in the method development in Chap. 4. This elaborateness is a by-product of trying to standardize a task into a teachable format. The intended take-home message of this book is actually the exact opposite. We believe the selected reaction monitoring experiment is so fundamentally sound, and the supporting resources so good, that rapid progress and success is achievable by any laboratory familiar with advanced mass spectrometry methods for proteins.

References

1. Thomson JJ (1913) Rays of positive electricity. Proc R Soc 89:1–20
2. Kinter M, Sherman NE (2000) Protein sequencing and identification using tandem mass spectrometry. Wiley, New York
3. Liu H, Sadygov RG, Yates JR 3rd (2004) A model for random sampling and estimation of relative protein abundance in shotgun proteomics. Anal Chem 76:4193–4201
4. Matzke MM, Brown JN, Gritsenko MA, Metz TO, Pounds JG, Rodland KD, Shukla AK, Smith RD, Waters KM, McDermott JE, Webb-Robertson BJ (2013) A comparative analysis of computational approaches to relative protein quantification using peptide peak intensities in label-free LC-MS proteomics experiments. Proteomics 13:493–503
5. Gygi SP, Rist B, Gerber SA, Turecek F, Gelb MH, Aebersold R (1999) Quantitative analysis of complex protein mixtures using isotope-coded affinity tags. Nat Biotechnol 17:994–999
6. Yao X, Freas A, Ramirez J, Demirev PA, Fenselau C (2001) Proteolytic 18O labeling for comparative proteomics: model studies with two serotypes of adenovirus. Anal Chem 73:2836–2842

7. Ross PL, Huang YN, Marchese JN, Williamson B, Parker K, Hattan S, Khainovski N, Pillai S, Dey S, Daniels S, Purkayastha S, Juhasz P, Martin S, Bartlet-Jones M, He F, Jacobson A, Pappin DJ (2004) Multiplexed protein quantitation in Saccharomyces cerevisiae using amine-reactive isobaric tagging reagents. Mol Cell Proteomics 3:1154–1169

8. Wühr M, Haas W, McAlister GC, Peshkin L, Rad R, Kirschner MW, Gygi SP (2012) Accurate multiplexed proteomics at the MS2 level using the complement reporter ion cluster. Anal Chem 84:9214–9221

9. Ong SE, Blagoev B, Kratchmarova I, Kristensen DB, Steen H, Pandey A, Mann M (2002) Stable isotope labeling by amino acids in cell culture, SILAC, as a simple and accurate approach to expression proteomics. Mol Cell Proteomics 1:376–386

10. Geiger T, Velic A, Macek B, Lundberg E, Kampf C, Nagaraj N, Uhlen M, Cox J, Mann M. (2013) Initial quantitative proteomic map of twenty-eight mouse tissues using the SILAC mouse. Mol Cell Proteomics 12:1709–1722

11. Washburn MP, Wolters D, Yates JR 3rd (2001) Large-scale analysis of the yeast proteome by multidimensional protein identification technology. Nat Biotechnol 19:242–247

12. Michalski A, Cox J, Mann M (2011) More than 100,000 detectable peptide species elute in single shotgun proteomics runs but the majority is inaccessible to data-dependent LC-MS/MS. J Proteome Res 10:1785–1793

13. Editors (2013) Method of the Year 2012. Nat Methods 10:1

14. Picotti P, Bodenmiller B, Aebersold R (2013) Proteomics meets the scientific method. Nat Methods 10:24–27

15. Gillette MA, Carr SA (2013) Quantitative analysis of peptides and proteins in biomedicine by targeted mass spectrometry. Nat Methods 10:28–34

16. Addona TA, Abbatiello SE, Schilling B, Skates SJ, Mani DR, Bunk DM, Spiegelman CH, Zimmerman LJ, Ham AJ, Keshishian H, Hall SC, Allen S, Blackman RK, Borchers CH, Buck C, Cardasis HL, Cusack MP, Dodder NG, Gibson BW, Held JM, Hiltke T, Jackson A, Johansen EB, Kinsinger CR, Li J, Mesri M, Neubert TA, Niles RK, Pulsipher TC, Ransohoff D, Rodriguez H, Rudnick PA, Smith D, Tabb DL, Tegeler TJ, Variyath AM, Vega-Montoto LJ, Wahlander A, Waldemarson S, Wang M, Whiteaker JR, Zhao L, Anderson NL, Fisher SJ, Liebler DC, Paulovich AG, Regnier FE, Tempst P, Carr SA (2009) Multi-site assessment of the precision and reproducibility of multiple reaction monitoring-based measurements of proteins in plasma. Nat Biotechnol 27:633–641

17. Beynon JH, Cooks RG, Amy JW, Baitinger WE, Ridley TY (1973) Design and performance of a mass-analyzed ion kinetic energy (MIKE) spectrometer. Anal Chem 45:1023A–1031A

18. McLafferty FW, Bente PF 3rd, Kornfeld R, Tsai S-C, Howe I (1973) Collision activation spectra of organic ions. J Am Chem Soc 95:2120–2129

19. Yost RA, Enke CG (1979) Triple quadrupole mass spectrometry for direct mixture analysis and structure elucidation. Anal Chem 51:1251–1264

20. Kruger TL, Litton JF, Kondrat RW, Cooks RG (1976) Mixture analysis by mass-analyzed ion kinetic energy spectrometry. Anal Chem 48:2113–2119

21. Kondrat RW, Cooks RG, McLaughlin JL (1978) Alkaloids in whole plant material: direct analysis by kinetic energy spectrometry. Science 199:978–980

22. Brotherton HO, Yost RA (1983) Determination of drugs in blood serum by mass spectrometry/mass spectrometry. Anal Chem 55:549–553

23. Johnson JV, Yost RA, Faull KF (1984) Tandem mass spectrometry for the trace determination of tryptolines in crude brain extracts. Anal Chem 56:1655–1661

24. Kalkum M, Lyon GJ, Chait BT (2003) Detection of secreted peptides by using hypothesis-driven multistage mass spectrometry. Proc Natl Acad Sci USA 100:2795–2800

25. Yi Z, Luo M, Carroll CA, Weintraub ST, Mandarino LJ (2005) Identification of phosphorylation sites in insulin receptor substrate-1 by hypothesis-driven high-performance liquid chromatography-electrospray ionization tandem mass spectrometry. Anal Chem 77:5693–5699

26. Roof RW, Haskell MD, Dukes BD, Sherman N, Kinter M, Parsons SJ (1998) Phosphotyrosine (p-Tyr)-dependent and -independent mechanisms of p190 RhoGAP-p120 RasGAP interaction: Tyr 1105 of p190, a substrate for c-Src, is the sole p-Tyr mediator of complex formation. Mol Cell Biol 18:7052–7063

27. Ruse CI, Willard B, Jin JP, Haas T, Kinter M, Bond M (2002) Quantitative dynamics of site-specific protein phosphorylation determined using liquid chromatography electrospray ionization mass spectrometry. Anal Chem 74:1658–1664
28. Willard BB, Ruse CI, Keightley JA, Bond M, Kinter M (2003) Site-specific quantitation of protein nitration using liquid chromatography/tandem mass spectrometry. Anal Chem 75:2370–2376
29. Arnott D, Kishiyama A, Luis EA, Ludlum SG, Marsters JC Jr, Stults JT (2002) Selective detection of membrane proteins without antibodies: a mass spectrometric version of the Western blot. Mol Cell Proteomics 1:148–156
30. Gerber SA, Rush J, Stemman O, Kirschner MW, Gygi SP (2003) Absolute quantification of proteins and phosphoproteins from cell lysates by tandem MS. Proc Natl Acad Sci USA 100:6940–6945
31. Barnidge DR, Goodmanson MK, Klee GG, Muddiman DC (2004) Absolute quantification of the model biomarker prostate-specific antigen in serum by LC-Ms/MS using protein cleavage and isotope dilution mass spectrometry. J Proteome Res 3:644–652
32. Anderson L, Hunter CL (2006) Quantitative mass spectrometric multiple reaction monitoring assays for major plasma proteins. Mol Cell Proteomics 5:573–588
33. Whiteaker JR, Zhang H, Zhao L, Wang P, Kelly-Spratt KS, Ivey RG, Piening BD, Feng LC, Kasarda E, Gurley KE, Eng JK, Chodosh LA, Kemp CJ, McIntosh MW, Paulovich AG (2007) Integrated pipeline for mass spectrometry-based discovery and confirmation of biomarkers demonstrated in a mouse model of breast cancer. J Proteome Res 6:3962–3975
34. Kinter M (2004) Toward a broader inclusion of liquid chromatography-mass spectrometry in the clinical laboratory. Clin Chem 50:1500–1502
35. MacLean B, Tomazela DM, Shulman N, Chambers M, Finney GL, Frewen B, Kern R, Tabb DL, Liebler DC, MacCoss MJ (2010) Skyline: an open source document editor for creating and analyzing targeted proteomics experiments. Bioinformatics 26:966–968
36. Desiere F, Deutsch EW, King NL, Nesvizhskii AI, Mallick P, Eng J, Chen S, Eddes J, Loevenich SN, Aebersold R (2006) The PeptideAtlas project. Nucleic Acids Res 34:D655–D658

Chapter 2
Specificity of Detection Is the Key Attribute of Selected Reaction Monitoring

Abstract Selected reaction monitoring uses tandem mass spectrometry to create instrumental conditions where only specific peptides can be detected. The two stages of the mass spectrometer are synchronized so that a signal is seen only when a molecular ion with a specific m/z is formed *and* that molecular ion fragments to a product ion with a specific m/z. The specificity of the analysis is increased by the unique amino acid sequence that defines a protein's identity and the effects of that unique amino acid sequence on the other parts of the overall assay, including the protein digestion and liquid chromatography. The result is the ability to selectively measure, with confidence and accuracy, any protein in a complex mixture.

2.1 Defining Selected Reaction Monitoring

The goal of this book is to describe a relatively new tool for measuring the amount of a protein in a biological system. This tool is selected reaction monitoring. The *reaction* in selected reaction monitoring is the collision induced dissociation of an ion formed in the electrospray ion source of a mass spectrometer. The tandem mass spectrometry experiment selects that ion, known as the precursor ion, in the first stage of m/z analysis, transmits those ions into a collision cell where the collision induced dissociation reaction occurs. Product ions formed by the fragmentation reaction are selected in the second stage of m/z analysis. Therefore, to produce a signal at the detector, a peptide must generate a molecular ion of the m/z selected by the first stage of mass analysis *and* fragment efficiently under the collision conditions being used to produce a fragment ion with the m/z selected in the second stage of mass analysis. This process is shown schematically in Fig. 2.1.

These experiments can be carried out in any type of tandem mass spectrometer, including instruments using either tandem in space or tandem in time arrangements, but the experiment is most often associated with triple quadrupole mass spectrometry

M. Kinter and C.S. Kinter, *Application of Selected Reaction Monitoring to Highly Multiplexed Targeted Quantitative Proteomics*, SpringerBriefs in Systems Biology, DOI 10.1007/978-1-4614-8666-4_2, © The Authors 2013

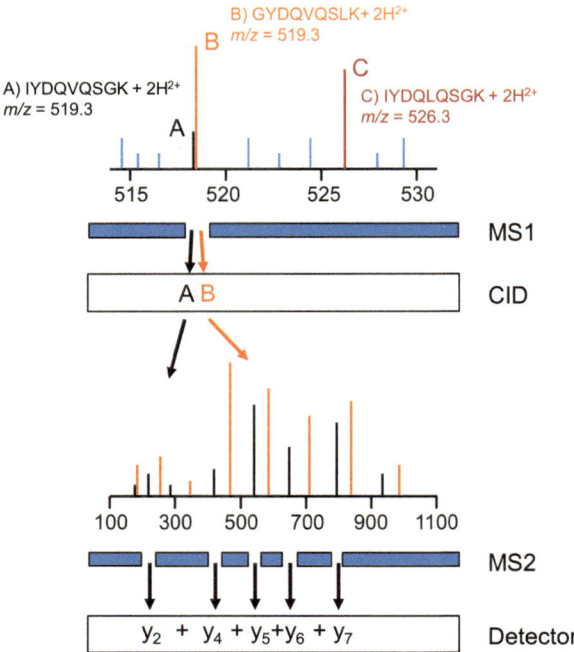

Fig. 2.1 A schematic illustration of the specificity of tandem mass spectrometry. A series of doubly charged peptide molecular ions are formed by electrospray ionization. The peptide ions are designated A, B, and C. Several ions of unknown origin are also illustrated. These ions may be other peptides, fragment ions of other peptides formed in the ion source, noise, plus many other possibilities. The first stage of mass analysis, MS1, transmits a single *m/z* of 519.3 which allows peptides A and B to enter the collision cell. All noise and peptide C are eliminated by MS1. In the collision cell, both molecular ions fragment into a series of product ions. The respective y-ion series are shown for each peptide. The second stage of mass analysis, MS2, steps between a predetermined set of *m/z*, allowing those fragment ions to reach the detector. In this illustration, an optimum set of y-ions from peptide A are selected and give a series of signals that are totaled to produce the overall signal for the peptide. Peptide B is eliminated in the second stage of mass analysis, so the entire signal recorded at this time is due to peptide A. This process can be completed in 25 ms or so, allowing up to approximately 40 peptides to be monitored in a 1 s cycle

systems. In most cases, selected reaction monitoring is used as a method to record chromatographic data in either liquid chromatography (LC) or gas chromatography (GC) experiments.

Some investigators use the term multiple reaction monitoring to reflect the nearly universal tendency to monitor more than one reaction in an experiment. The term selected reaction monitoring is the preferred term that will used here. Based on the International Union of Pure and Applied Chemistry the term multiple reaction monitoring is deprecated [1]. Abbreviations for this experiment include SRM and MRM, but we will not use these abbreviations here believing that fewer abbreviations will enhance the clarity of our presentation. A short list of the abbreviations that we do use is given in the front material.

2.2 Specificity of Analysis Is the Key Attribute

In the early 1980s, as the field of tandem mass spectrometry was beginning to flourish, McLafferty described the 'four Ss' of good trace analyses—sensitivity, selectivity, speed, and $ [2]. A continual point made through a series of papers published by McLafferty and the other pioneers in the field was that tandem mass spectrometry offered the potential for a remarkable increase in selectivity of detection that could have a dramatic influence on the effectiveness of the analytical method—increasing the accuracy, lowering the limit of detection, increasing the speed, and lowering the cost. Therefore, understanding the concept of selectivity, and the role it plays in an analytical method, is crucial to understanding the strengths of measuring protein abundance by selected reaction monitoring.

By definition, quantitative analyses are designed to determine the amount of a certain chemical entity (the analyte) in a sample. To accomplish this goal, quantitative methods measure some parameter and relate that measurement to the amount of the analyte. Thus, a key assumption in all quantitative methods is that the parameter being measured is due to the analyte and only the analyte. Other compounds that also contribute to the measurement are defined as interferences. The number and magnitude of these interferences can vary greatly. Severe interferences can render an assay unusably inaccurate. In other cases, more moderate levels of interference will require some element of compensation to maintain accuracy. Finally, some interferences can be so minor that the assay remains accurate and useful.

There are many approaches to increasing the selectivity of an assay. For example, steps can be added to the sample processing that remove the interferences from the final sample that is analyzed. A related strategy, if the analysis uses chromatography as a component, would be to increase the resolving power of the chromatography with tools like longer columns, specialized columns, gradient elution, ultrahigh pressure, or a combination. However, the most fundamental tool to minimize the potential for interferences in an assay is to increase the selectivity of the parameter being measured. In other words, measure a parameter that is as unique as possible for the analyte of interest, thereby discriminating against all other components of the sample.

Selected reaction monitoring is an example of measuring a parameter that is as unique as possible. The uniqueness of the parameter measured in selected reaction monitoring is derived from the two linked stages of m/z analysis, also called mass analysis. To generate a signal, the chemical entity *must both* (a) produce an ion with the m/z selected by the first mass analyzer and (b) fragment to produce a product ion with the m/z selected by second mass analyzer. Yost described this process as having 'high information content' [3]. This high information content reflects the fact that each stage of mass analysis is highly discriminating and linking the two stages defines an addition chemical characteristic.

2.3 The Specificity of Selected Reaction Monitoring Is Ideal for Targeted Quantitative Proteomics

The effectiveness of selected reaction monitoring in the analysis of proteins has three major elements (1) the uniqueness of a protein's amino acid sequence, (2) the conversion of the protein to peptides that preserve the amino acid sequence information, and (3) the ability of an LC-tandem mass spectrometry experiment to use the unique properties derived from a peptide's amino acid sequence.

The amino acid sequence of a protein is a fundamental attribute that distinguishes that protein from all other proteins. Although many proteins have similar isoforms and transcript variants, all differ in some way at the amino acid sequence level. One should also appreciate that while the proteome of any species is large, it is finite and well-defined. Analysis of the human genome indicates approximately 22,000 human genes and the human RefSeq database contains approximately 32,000 known protein coding transcripts [4]. Digestion of these proteins with a protease such as trypsin, an inherent part of the mass spectrometry experiment, translates these unique amino acids sequences into a large set of unique peptide sequences that can be traced back to the parent protein. This link between a peptide sequence and the parent protein is the foundation of the protein identification process, whether the peptide amino acid sequence is determined by tandem mass spectrometry or its predecessor Edman degradation [5]. Essentially all proteins, when digested with trypsin, will produce at least one and most likely many peptides with unique amino acid sequences. Selectively detecting these peptides serves as the critical marker for the parent protein both qualitatively as the protein of interest and quantitatively showing how much of the protein is in the sample.

The amino acid sequence gives each peptide a distinctive set of chemical characteristics that determine its behavior in the LC-tandem mass spectrometry experiment. These chemical characteristics are the hydrophobicity of the peptide, number of basic amine groups, molecular weight, and fragmentation pattern (in no particular order). Differences in hydrophobicity are reflected in the chromatographic retention and the response in electrospray ionization, which affect the detectability of a peptide. The number of basic amine groups determines the favored charge state. Finally, the most important factors, molecular weight and fragmentation pattern, are utilized by the tandem mass spectrometer and constitute the reaction monitored in the selected reaction monitoring experiment.

2.4 Turning the Specificity of Selected Reaction Monitoring into a Functioning Assay

The remainder of this book describes how our lab uses these elements to build and execute selected reaction monitoring experiments with sufficient specificity to accurately and precisely measure the amounts of multiple proteins in complex samples. The assays that are develop take advantage of all contributors to specificity.

The process begins with careful selection of the peptides to be measured and the tandem mass spectrometry conditions needed to carry out those measurements, including which peptides to use, which fragmentation reactions to monitor, and the collision energy needed to maximize the efficiency of those reactions. The design process is described in Chap. 3. As stated in that description, the method is presented as a methodical multistep process that appears tedious and time consuming—it is neither. In fact, a possibly unappreciated part of the design process is that most proteins have a good number of suitable peptides to utilize. The sample preparation method described in Chap. 4 also contributes. We use a short-run SDS-PAGE method that provides an effective clean-up of the sample through the protein-specific migration into the gel and subsequent washing of any remaining small molecule contaminants. Finally, as described in Chap. 5, the chromatographic separation of the peptides is used to create a modest time window in which each peptide is monitored. This scheduling based on each peptide's specific retention time improves the recognition of the correct signal by ignoring other signals that might otherwise generate confusion. Scheduling also allows the assembly of multiple protein-specific methods into a multiplexed assay. Taken together, sample preparation, chromatographic separation, plus the high specificity of selected reaction monitoring creates the strong link between the parameter being measured and the analyte of interest. This strong link, and the resulted accuracy and precision, is the fundamental power of targeted quantitative proteomics.

References

1. IUPAC (nd) Standard definitions of terms relating to mass spectrometry. Physical and Biophysical Chemistry Division. International Union of Pure and Applied Chemistry. http://mass-spec.lsu.edu/msterms/index.php/Selected_reaction_monitoring
2. McLafferty FW (1981) Tandem mass spectrometry in trace toxicant analysis. Biomed Mass Spectrom 8:446–448
3. Yost RA, Fetterolf DD (1983) Tandem mass spectrometry (MS/MS) instrumentation. Mass Spectrom Rev 2:1–45
4. Pruitt KD, Tatusova T, Brown GR, Maglott DR (2012) NCBI reference sequences (RefSeq): current status, new features and genome annotation policy. Nucleic Acids Res 40:D130–135
5. Kinter M, Sherman NE (2000) Protein sequencing and identification using tandem mass spectrometry. Wiley, New York

Chapter 3
Designing a Selected Reaction Monitoring Method

Abstract The instrument instructions for a selected reaction monitoring experiment contain the basic programming needed to detect the set of peptides that will act as the quantitative surrogate for each protein being measured. The overarching goal of the design process is find the best peptides to monitor and optimize the collision induced dissociation conditions to give the best signal for those peptides. The general stages are finding an initial set of candidate peptides for initial testing, optimizing the conditions for those candidates, selecting the best of the optimized group, and validating the proper detection of the correct peptides. A six step approach is used to go through these stages.

3.1 Overview of the Design Process

As described in Chap. 2, selected reaction monitoring is an instrumental technique in which a tandem mass spectrometer is used to detect, with high specificity, an analyte (in this case a peptide). This detection is based on the peptide producing a molecular ion with a characteristic m/z *and* fragmenting by collision induced dissociation to a set of product ions that have a characteristic set of m/z. This process is illustrated in Fig. 3.1. Our laboratory uses the term *descriptor* as a name for the set of instructions used by the tandem mass spectrometer for the coordinated selection of parent ion m/z in the first stage of mass analysis and product ion m/z in the second stage of mass analysis, with predetermined and optimized collision induced dissociation conditions. The assay for each protein is composed of detecting multiple peptides, with the detection of each peptide based on monitoring multiple collision induced dissociation reactions. The overall assay is built by combining descriptors for multiple proteins. This chapter will describe the strategy used to develop these descriptors.

As a starting point, it is important to appreciate the factors that make the selected reaction monitoring experiment work for measuring the amount of a protein.

M. Kinter and C.S. Kinter, *Application of Selected Reaction Monitoring to Highly Multiplexed Targeted Quantitative Proteomics*, SpringerBriefs in Systems Biology, DOI 10.1007/978-1-4614-8666-4_3, © The Authors 2013

Fig. 3.1 An overview of reactions in selected reaction monitoring. The process is exemplified with a tryptic peptide from ACOX1. Only a subset of the fragmentation reactions are illustrated for clarity. The details of these reactions are used to construct the descriptor that communicates the conditions needed to detect these peptides to the LC-tandem mass spectrometry system

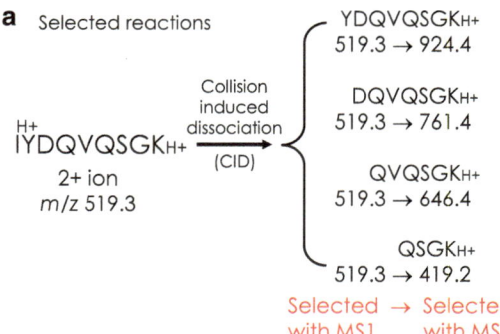

a Selected reactions

YDQVQSGK$_{H+}$
519.3 → 924.4

Collision
induced
dissociation

H+
IYDQVQSGK$_{H+}$ → (CID)

2+ ion
m/z 519.3

DQVQSGK$_{H+}$
519.3 → 761.4

QVQSGK$_{H+}$
519.3 → 646.4

QSGK$_{H+}$
519.3 → 419.2

Selected → Selected
with MS1   with MS2

b Example descriptor

Parent ion m/z	Product ion m/z	Collision energy, eV	Start, minutes	End, minutes
519.27	924.44	18	11.5	14.5
519.27	761.38	14	11.5	14.5
519.27	646.35	16	11.5	14.5
519.27	419.22	18	11.5	14.5

We see four fundamental components to consider. *First*, the primary amino acid sequence of a protein is static and known via the various genome sequencing efforts. Although a myriad of splice variants, point mutants, post-translational processing, post-translational modification, etc. may create a large numbers of variants for any protein, the core amino acid sequence remains the same. *Second*, the enzyme trypsin is an effective tool for protein digestion. As discussed in Chap. 4, the combination of good denaturation, reduction and alkylation, and the effectiveness of proteomics-grade trypsin mean that a reproducible set of peptides is produced from this primary amino acid sequence in high yield. *Third*, the elution and ionization of the peptides in a reversed phase LC-electrospray ionization tandem mass spectrometry system is determined by the chemistry of those peptides, making their response similarly reproducible. *Fourth*, the details of the collision induced dissociation of peptide ions are determined by their amino acid sequence, making the fragmentation pattern reproducible. This reproducibility includes both the *m/z* of the product ions that are seen and their relative abundance.

The essence of the descriptor design process is to choose from a set of candidates, the peptides, product ions, and collision induced dissociation conditions that give the best response. While the description of this process may seem tedious, it is not difficult because most proteins will produce a surplus of effective peptides. Once established, a well-designed selected reaction monitoring descriptor should not require extensive redesign because the amino acid sequence, specificity of trypsin, chemistry of the peptides, and basic chemistry of reversed phase liquid chromatography will not change. The overall LC-tandem mass spectrometry instrument will certainly change with time but one would expect this change should require only a reestablishment of the retention characteristics and collision induced dissociation conditions for the selected peptides.

3.2 Basic Peptide Fragmentation

The factors that determine the gas phase ion chemistry of peptide fragmentation reactions are still being examined by a number of investigators [1–5]. Several basic principles, however, have been established and an understanding of those principles will help recognize characteristics of more useful versus less useful peptides. Figure 3.2 summarizes the basic steps involved in the collision induced dissociation of a peptide ion formed by electrospray ionization. This process is also discussed in greater detail from the standpoint of interpreting a spectrum in another book [6].

As a starting point, the enzyme trypsin specifically hydrolyzes the amide bond at the C-terminal side of lysine and arginine residues, except when the adjacent amino acid is a proline. This specificity means that most tryptic peptides will have only two strong basic sites, the N-terminus and the side chain of the C-terminal lysine or arginine. With electrospray ionization, the peptide becomes protonated at these sites, with one proton at the N-terminus and the other proton at the C-terminus. This protonation gives a doubly charged molecular ion with generally minor amounts of singly or triply charged molecular ions. With collision induced dissociation, the kinetic energy of the collision is converted to internal energy in the peptide ion that drives the subsequent steps. Specifically, the N-terminal proton becomes mobile, migrates to the different amide bonds, and directs the breaking of that bond. Because of the location of the protons, this fragmentation reaction produces two complementary fragment ions. The y-ions are stable product ions that contain the C-terminus of the peptide. The b-ions contain the N-terminus of the peptide and may not be completely stable. This instability allows the larger b-ions to fragment further under multiple collision conditions until the b_2-ion is formed.

The CID spectrum of a peptide is a histogram that shows the frequency of each fragmentation reaction as relative abundance. Product ions with the highest relative abundance are formed by the fragmentation reactions that happen most often. These favored reactions will be influenced by a variety of factors but a primary factor is the nature of the amino acid at that position. Certain amino acids will direct fragmentation to a favored position to generate higher relative abundance product ions. By definition, the presence of favorable fragmentation sites with higher relative abundance ions must be accompanied by unfavorable sites giving lower relative abundance product ions. A number of other fragmentation reactions will also occur and contribute to the

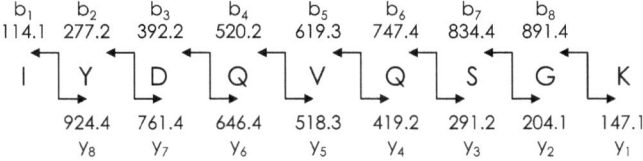

Fig. 3.2 Peptide fragmentation at the amide bonds. Since protons are present and remain on each product of the fragmentation reaction, two complementarily series of ions are formed. The y-ions contain the C-terminus and the b-ions contain the N-terminus. The structure of the b-ions prevents the formation of a b_1-ion. Not shown in this figure is the formation of the a-ions. The primary a-ion seen is the a_2-ion

spectrum. The most common reactions are the loss of ammonia and water. The results of these reactions are seen as additional ions in the product ion spectrum at -17 Da and -18 Da from the fragment ion losing the ammonia or water. In some cases, abundant ions from the loss of ammonia and water by the molecular ion will also be seen. In the case of doubly charged molecular ions, these fragment ions will appear as doubly charged ions at approximately -9 Da, -18 Da, and -27 Da from the m/z of the molecular ion being fragmented, depending on the number of losses. In general, a strong tendency to lose water and ammonia from either the molecular ion or the product ions is an undesirable attribute because product ion signal is lost in these many competing reactions.

In the design of a selected reaction monitoring descriptor, the information in the CID spectrum is an important starting point. The general approach is to seek out peptides for which the majority of the abundance in the molecular ion is contained in a modest number of product ions. The product ion spectra of the peptides observed for a protein facilitate this selection process. As will be seen, however, the final selection and the refinement of conditions are based on actual experimental testing and are optimized to maximize the utility of the final set of peptides that are used.

3.3 Background Resources for the Design Process

The design of selected reaction monitoring descriptors is aided by several internet resources. As a group, these resources allow one to retrieve the amino acid sequence of the protein and any related isoforms, evaluate the alignment of the sequences from multiple isoforms and across species, examine the peptides detected for a protein and their product ion spectra, and calculate predicted chromatographic retention times. All these tools can be used to directly aid the design process. Finally, other tools are available to estimate the level of expression and help select the most useful tissue in which to carry out the design optimization experiments. The resources we use are the BioGPS, the PeptideAtlas, Clustal Omega, and the Sequence Specific Retention Calculator. Other similar resources are also available, but these are the ones we use and are most able to discuss.

The BioGPS is a gene annotation portal that allows users to browse an array of information for a given protein [7]. In the descriptor design process, the resource is used to find two key pieces of information, the existence of multiple isoforms of the protein and a general sense of the expression at the message level in various tissues. The potential for multiple isoforms is seen in the number of protein accession numbers seen in the gene identifiers portion of the search result. Accession numbers beginning with NP designate entries in RefSeq database and each entry refers to a different isoform. It is important to consider these isoforms during the descriptor development process because selected peptides may not be seen in all isoforms. Some proteins will have only a single isoform and others will have several isoforms that are nearly identical. The Clustal Omega tool described below is used to align and compare isoforms. The decision to include or not include different isoforms depends of the goals of the experiment. The other key piece of information seen in the BioGPS system is tissue-specific expression information. In our experience, this expression information is a key part of starting the design process for any protein. These data are based on Affymetrix gene chip experiments accessed by the BioGPS through various publically available datasets. The data are shown based on the gene name for each Affymetrix probe for each analyzed tissue. The goal when accessing these data is to judge which tissues will have low, medium, or high levels of expression for the protein being studied. Although the data are given in Affymetrix expression units, it is relatively straight-forward to develop sufficient experience to understand the magnitude of the numbers and judge the relative protein expression.

The PeptideAtlas is an internet resource that contains large sets of mass spectrometry data for the identification of different proteins [8]. These data are contributed by a group of leading proteomics laboratories. Extensive amounts of data are available for human, mouse, and yeast experiments. The information in this database helps the design process in two important ways. *First,* the peptides that have been detected by these laboratories for a specified protein can be sorted based on the number of times observed in the contributed work. In general, the more detectable peptides will have been seen most often. Therefore, this frequency allows one to begin the design process based on an expectation of which peptides are the best candidates for the analysis. *Second*, representative CID spectra of those peptides can be retrieved and evaluated. As noted above, seeing these spectra allow one to judge the complexity of the fragmentation pattern with a desire to identify peptides with less complex patterns.

The final two Internet-based programs that can be used to facilitate the design process are Clustal Omega and the Sequence Specific Retention Calculator (SSRC). The Clustal Omega program aligns multiple protein sequences [9]. These alignments are used to evaluate different isoforms, if applicable, when selecting peptides. It may also be useful to align the sequences from multiple species. This alignment is certainly important if designing descriptors that might be used for samples from different species. Aligning multiple species can also help in cases where more mass spectrometry data is available in the PeptideAtlas for another species. In many cases, information about homologous peptides can help the design when no other direct information can be found. The Sequence Specific Retention Calculator calculates a hydrophobicity value of a given peptide sequence based on the amino acid content [10]. This hydrophobicity value can then be used to calculate a predicted retention times for new peptides based on observed retention times for a set of references peptides on the LC system being used. In some cases, when multiple usable peptides are available, it may be advisable to avoid peptides that do not seem to elute at the predicted retention time.

3.4 An Example of Descriptor Development Based on ACOX1

Acyl-Coenzyme A oxidase 1, palmitoyl (ACOX1) is a peroxisomal enzyme with a molecular weight of 74 kDa that catalyzes the first reaction in the beta oxidation of fatty acids. An interesting part of the function of this enzyme is the fact that it uses oxygen as its electron acceptor to produce hydrogen peroxide directly in this reaction. ACOX1 will be used as an example for the descriptor development process. The example will refer to experiments in mouse heart, liver, and skeletal muscle. The development process has six basic steps, with each described in greater detail in the following sections:

1. Assemble background information about the protein, including the amino acid sequence and information about the level of expression.
2. Select candidate peptides for the initial descriptor.
3. Test the candidate peptides experimentally to determine the best performing peptides and their retention times.
4. Optimize the collision energy for each fragmentation reaction.
5. Select the best fragmentation reactions for the best peptides for the final descriptor.
6. Test and validate the final descriptor.

(1) Assemble background information about the amino acid sequence and expression

At the beginning of the design process, it is important to have a basic understanding of the target protein sequence, any variants in that sequence, and what level of expression to expect in the tissues in which it will be measured. This initial step is facilitated by use of the BioGPS resource and the Clustal Omega alignment program. The initial search of Acox1 with BioGPS gives two important pieces of information for the design process—information about the number of isoforms based on multiple entries in the RefSeq protein sequence database and a general expectation for protein abundance based on message levels in the summary of Affymetrix gene chip experiments.

Two isoforms of the mouse protein exist, isoform 1 and 2 (accession numbers NP_056544 and NP_001258827, respectively). The details of the different isoform are given in the database entries which note that both isoforms have the same length but have numerous differences in the coding region. Alignment of the two mouse protein sequences with Clustal Omega shows 96 % overall identity and 100 % identity in all parts of the protein except in the region of amino acids 90–130, which have very low identity between the isoforms (Fig. 3.3). Further, alignment between the mouse and human sequences (with 3 very similar isoforms) also showed high identity. Although this amount of similarity would generally indicate a good possibility to design a descriptor that could be used for both the mouse and human protein, this is surprisingly difficult in this case. Many of the leading mouse candidates will not be seen in human samples because of differences at trypsin cleavage sites.

The second piece of background information from the BioGPS search is that the expression of the protein in the mouse is relatively high in the liver and significantly less in heart and skeletal muscle. These data are shown in Table 3.1. ACOX1 message is detected via three probes, with two of the probes having values that are much higher than the third probe. For the purpose of predicting expression, we focus on the maximum values believing that these probes give the best response for that gene.

```
mouse 1    MNPDLRKERAAATFNPELITHILDGSPENTRRRREIENLILNDPDFQHEDYNFLTRSQRYEVAVKKSATM
mouse 2    MNPDLRKERAAATFNPELITHILDGSPENTRRRREIENLILNDPDFQHEDYNFLTRSQRYEVAVKKSATM
human A    MNPDLRRERDSASFNPELLTHILDGSPEKTRRRREIENMILNDPDFQHEDLNFLTRSQRYEVAVRKSAIM
           ****** : **  : * : **** : ********* : ********* : ********** ************** : *** *

mouse 1    VKKMREFGIADPEEIMWFKNSVHRGHPEPLDLHLGMFLPTLLHQATEEQQERFFMPAWNLEITGTYAQTE
mouse 2    VKKMREFGIADPEEIMWFKKLHMVNFVEPVGLNYSMFIPTLLNQGTTAQQEKWMHPSQELQIIGTYAQTE
human A    VKKMREFGIADPDEIMWFKNFVHRGRPEPLDLHLGMFLPTLLHQATAEQQERFFMPAWNLEIIGTYAQTE
           *********** : ****** :  ** : *.  . ** : **** . * . *  ***    *** : : :  * :  : * : * *******

mouse 1    MGHGTHLRGLETTATYDPKTQEFILNSPTVTSIKWWPGGLGKTSNHAIVLAQLITRGECYGLHAFVVPIR
mouse 2    MGHGTHLRGLETTATYDPKTQEFILNSPTVTSIKWWPGGLGKTSNHAIVLAQLITRGECYGLHAFVVPIR
human A    MGHGTHLRGLETTATYDPETQEFILNSPTVTSIKWWPGGLGKTSNHAIVLAQLITKGKCYGLHAFIVPIR
           ******************* : ********************************* : * : ****** : ****

mouse 1    EIGTHKPLPGITVGDIGPKFGYEEMDNGYLKMDNYRIPRENMLMKYAQVKPDGTYVKPLSNKLTYGTMVF
mouse 2    EIGTHKPLPGITVGDIGPKFGYEEMDNGYLKMDNYRIPRENMLMKYAQVKPDGTYVKPLSNKLTYGTMVF
human A    EIGTHKPLPGITVGDIGPKFGYDEIDNGYLKMDNHRIPRENMLMKYAQVKPDGTYVKPLSNKLTYGTMVF
           ********************** : * : ********** : *************************************

mouse 1    VRSFLVGSAAQSLSKACTIAIRYSAVRRQSEIKRSEPEPQILDFQTQQYKLFPLLATAYAFHFLGRYIKE
mouse 2    VRSFLVGSAAQSLSKACTIAIRYSAVRRQSEIKRSEPEPQILDFQTQQYKLFPLLATAYAFHFLGRYIKE
human A    VRSFLVGEAARALSKACTIAIRYSAVRHQSEIKPGEPEPQILDFQTQQYKLFPLLATAYAFQFVGAYMKE
           ******. ** : : ***************** :  **** : ** : ********************* : * : : *****

mouse 1    TYMRINESIGQGDLSELPELHALTAGLKAFTTWTANAGIEECRMACGGHGYSHSSGIPNIYVTFTPACTF
mouse 2    TYMRINESIGQGDLSELPELHALTAGLKAFTTWTANAGIEECRMACGGHGYSHSSGIPNIYVTFTPACTF
human A    TYHRINEGIGQGDLSELPELHALTAGLKAFTSWTANTGIEACRMACGGHGYSHCSGLPNIYVNFTPSCTF
           ** **** . ********************** : **** : *** : **** ******** : * . *** . *** . ***

mouse 1    EGENTVMMLQTARFLMKIYDQVQSGKLVGGMVSYLNDLPSQRIQPQQVAVWPTLVDINSLDSLTEAYKLR
mouse 2    EGENTVMMLQTARFLMKIYDQVQSGKLVGGMVSYLNDLPSQRIQPQQVAVWPTLVDINSLDSLTEAYKLR
human A    EGENTVMMLQTARFLMKSYDQVHSGKLVCGMVSYLNDLPSQRIQPQQVAVWPTMVDINSPESLTEAYKLR
           ***************** : **** : *** : * . ***************************** : ***** : * . ********

mouse 1    AARLVEIAAKNLQAQVSHRKSKEVAWNLTSVDLVRASEAHCHYVTVKVFADKLPKIQDRAVQAVLRNLCL
mouse 2    AARLVEIAAKNLQAQVSHRKSKEVAWNLTSVDLVRASEAHCHYVTVKVFADKLPKIQDRAVQAVLRNLCL
human A    AARLVEIAAKNLQKEVIHRKSKEVAWNLTSVDLVRASEAHCHYVVVKLFSEKLLKIQDKAIQAVLRSLCL
           *********** : * : * : *** . ** : ******************* . * : : ** ****** : * . ********

mouse 1    LYSLYGISQKGGDFLEGNIITGAQMSQVNSRILELLTVTRPNAVALVDAFDFKDVTLGSVLGRYDGNVYE
mouse 2    LYSLYGISQKGGDFLEGNIITGAQMSQVNSRILELLTVTRPNAVALVDAFDFKDVTLGSVLGRYDGNVYE
human A    LYSLYGISQNAGDFLQGSIMTEPQITQVNQRVKELLTLIRSDAVALVDAFDFQDVTLGSVLGRYDGNVYE
           ********* : . **** : * . * . * : * : . ****  : ** : ********** : ****************** :***

mouse 1    NLFEWAKKSPLNKTEVHESYYKHLKPLQSKL
mouse 2    NLFEWAKKSPLNKTEVHESYYKHLKPLQSKL
human A    NLFEWAKNSPLNKAEVHESY-KHLKSLQSKL
           ****** : ***** : ****** **** ******
```

Fig. 3.3 Alignment of ACOX1 sequences. The two mouse isoforms of ACOX1, isoform 1 and isoform 2, are aligned with one of three human isoforms, isofrom a. For the mouse proteins, the two isoforms have large regions of identical sequence. This identity makes it straight-forward to select peptides that are good for the detection of both isoforms. While the human isoform also has high identity, it is notable that none of the three peptides used in the mouse ACOX1 assay are seen in the human protein due to changes at tryptic sites

Table 3.1 Summary of ACOX1 Affymetrix expression values in BioGPS for selected tissues

Affymetrix probe number	Liver	Skeletal muscle	Heart
1416408_at	20,371	1,271	2,420
1416409_at	17,066	932	1,920
1444518_at	324	5	14

In the liver, ACOX1 message expression is in the range of 20,000 units in the Affymetrix system. In our experience, mouse proteins with values greater than 10,000 units are usually readily detectable with a well-designed selected reaction monitoring assay. As a result, these tissues are also ideal for the design process because the peptide detection is readily recognizable. In contrast, proteins with expression less than 5,000 units, as seen for both skeletal muscle and heart, can be more challenging to detect in the selected reaction monitoring assay, making the design process challenging if limited to those samples. The overall expectation, based on these data, would be that the design process is best carried out with liver samples but, once designed, one should expect to be able to detect ACOX1 in both heart and skeletal muscle but with significantly lower signals.

(2) Select candidate peptides for the initial descriptor

As discussed above, the details of the collision induced dissociation of peptide ions are determined by their amino acid sequence, making the fragmentation pattern reproducible. This reproducibility includes both the *m/z* of the product ions that are seen and their relative abundance. This reproducibility, while not absolute, extends to some degree across laboratories and instrument systems. Therefore, collections of mass spectrometry data from multiple laboratories are a valuable resource for the descriptor design process. Further, these databases have grown in size to the point that they are an acceptable replacement for any local experience with a give protein for this design process.

The first use of the PeptideAtlas is to see the combined experience of the multiple laboratories for the coverage of the protein sequence. As of April 2013, the laboratories contributing to PeptideAtlas Mouse Build 2013-2 had 3,881 observations of ACOX1 peptides covering 79 % of the protein sequence. These peptides can be sorted based on the number of observations to provide a measure of the detectability of each peptide. As is the basis of label-free quantitative methods, data-dependent analyses tend to detect the peptides giving the strongest signals more often. Table 3.2 shows the top 30 peptides in the sorted PeptideAtlas list for ACOX1. Building the descriptor begins with a simple evaluation of the list of peptides seen for the protein to find the peptides that are most likely to give the best results. This list can come from the PeptideAtlas, as in the case of this ACOX1 example, or from local data.

A concept described by Aebersold is the *best flyer* concept [11]. The idea of the best flyer is that any group of proteins will produce a corresponding group of peptides with a continuum of responses in reversed phase LC-electrospray ionization tandem mass spectrometry experiment. The best responding peptides in this continuum will give a maximal response, defined as amount of signal per amount of peptide, which is independent of the protein they come from. As discussed in greater detail in Chap. 5, the utility of the best flyer model is that it allows relative protein abundances to be estimated directly from the signal strength.

The best flyer concept also captures the simple goal of the descriptor design process for each protein—to identify and use the peptides that give the strongest signal. Our consideration of the different possible peptide begins with the ideal that all peptides have the potential to be best flyers, but different negative factors in the amino acid sequence take this potential away. One should remember that these negative factors include not only the mass spectrometry but also the chemical steps associated with the sample preparation and the liquid chromatography component

Table 3.2 Mouse ACOX1 in the PeptideAtlas

	Accession	Pre amino acid	Sequence	Empirical suitability score (ESS)	Number of observations
1	PAp00184451	K	FGYEEMDNGYLK	1	308
2	PAp00184782	K	TSNHAIVLAQLITR	0.87	230
3	PAp00184983	K	GGDFLEGNIITGAQMSQVNSR	0.86	222
4	PAp00184963	K	LVGGMVSYLNDLPSQR	0.86	221
5	PAp00185196	R	EIGTHKPLPGITVGDIGPK	0.85	213
6	PAp00183946	K	**TQEFILNSPTVTSIK**	0.82	196
7	PAp00396347	K	LFPLLATAYAFHFLGR	0.81	192
8	PAp00184671	R	INESIGQGDLSELPELHALTAGLK	0.8	183
9	PAp00184177	R	ILELLTVTRPNAVALVDAFDFK	0.8	182
10	PAp00184375	R	**GLETTATYDPK**	0.78	175
11	PAp00184709	K	YAQVKPDGTYVKPLSNK	0.72	136
12	PAp00184515	R	**YDGNVYENLFEWAK**	0.7	121
13	PAp00185015	R	NLCLLYSLYGISQK	0.7	121
14	PAp00350462	K	**EVAWNLTSVDLVR**	0.69	116
15	PAp00184433	R	**SFLVGSAAQSLSK**	0.67	104
16	PAp00184669	R	AAATFNPELITHILDGSPENTR	0.65	93
17	PAp00184787	R	GECYGLHAFVVPIR	0.65	93
18	PAp00396807	K	NLQAQVSHR	0.64	85
19	PAp00399124	K	MREFGIADPEEIMWFK	0.42 [mc]	82
20	PAp00395910	K	AFTTWTANAGIEECR	0.63	77
21	PAp00184972	R	SEPEPQILDFQTQQYK	0.61	65
22	PAp00184610	K	**DVTLGSVLGR**	0.6	64
23	PAp00399021	K	RSEPEPQILDFQTQQYK	0.40 [mc]	56
24	PAp00396592	K	SKEVAWNLTSVDLVR	0.39 [mc]	54
25	PAp00395946	K	LTYGTMVFVR	0.58	52
26	PAp00397265	R	ASEAHCHYVTVK	0.58	48
27	PAp00398843	K	**IYDQVQSGK**	0.56	39
28	PAp00185502	R	EFGIADPEEIMWFK	0.56	34
29	PAp00396874	R	AVQAVLR	0.05 [mgl]	28
30	PAp00185049	R	IQPQQVAVWPTLVDINSLDSLTEAYK	0.54	25

Partial list from a total of 61 peptides that have at least 1 observation in the database. The accession number is a unique identified in the PeptideAtlase database. The Empirical Suitability Score (ESS) is a calculation designed to predict the best peptides to use in a selected reaction monitoring experiment

of the analysis. A common concern is the variability of these processes for certain peptides based on these steps. Many of these factors are recognizable, but some must be tested experimentally.

In our laboratory, three peptide amino acid sequence features quickly disqualify a peptide from further consideration. Methionine (M)-containing peptides are not used because of variable oxidation during the gel electrophoresis. While these peptides may be useful in samples that are digested without SDS-PAGE, the ease of this oxidation would still be a concern. In the case of ACOX1, three of the four most often observed peptides contain methionine and are eliminated from consideration. Peptides with a missed cleavage are not used because of the variability of the trypsin reaction at that site. While the missed cleavage peptides may be observed frequently, it is likely that the full tryptic peptide is also formed to some degree. The PeptideAtlas designates these peptides with the [mc] designation as seen for three peptides on this list. Finally, peptides with an N-terminal glutamine (Q) are not used because they can cyclize to a variable degree to form pyroglutamate. No such peptides are seen in the case of ACOX1.

A less absolute disqualifying feature is an additional basic site in the peptide. These sites are produced by either histidine (H) residues or lysine and arginine residues followed by a proline (P)—the K–P and R–P bonds not cleaved by trypsin. These amino acids can have two unwanted effects. The most harmful unwanted effect is the formation of both doubly and triply charged molecular ions which divides the available peptide ions into two distinctly different m/z-forms. The second, less predictable unwanted effect is the ineffective fragmentation of these peptides. In some cases, peptides containing histidine, K–P, and R–P sites may be useful if they occur within two amino acids of the N- or C-terminii. In these cases, the proximity of the basic sites will tend limit the abundance of the triply charged ion. These peptides will often be eliminated at the testing stage but, depending on the number of other possibilities, it may be prudent to include in this first candidate-building stage.

Another less absolute disqualifying feature is the presence of a cysteine (C) residue in a peptide. These peptides may be more variable because of any variability in the reduction and alkylation reactions. It has been our experience, however, that these reactions are very effective and have reproducibly high yield. Therefore, we do not automatically eliminate cysteine-containing peptides from consideration.

Table 3.2 also denotes a peptide, AVQAVLR, which is seen in other protein sequences. Blast searches can be used to investigate this overlap further if needed. In this case, the peptides sequence is seen as a half-tryptic peptide in ATP-binding cassette sub-family F member 1 (ABCF1). This type of overlap could affect the specificity of the analysis for ACOX1.

Finally, other parameters such as the molecular weight of the peptide or peptides that might have poor reversed-phase liquid chromatography characteristics might also be avoided. The molecular weight limit not only reflects the m/z limits of the mass spectrometry system, but also the tendency of larger peptides to produce a more extensive group of fragment ions, which will require monitoring correspondingly more reactions. While not absolutely disqualifying, this issue may make longer peptides less useful in a practical sense. The poor liquid chromatography characteristics to be avoided are peptides that elute either too early or too late. Although not included in Table 3.2, the PeptideAltas includes a calculated measure of hydrophobicity that can be used to recognize such peptides.

As this initial elimination process is completed, it should be noted that a relatively large protein like ACOX1 will typically present an excess of seemingly ideal peptides for the selected reaction monitoring descriptor. In our experience, this abundance of choices is common with the main exceptions being small (less than 20 kDa) proteins that give fewer possible peptides.

Table 3.3 Retention time and intensity for the initial analysis of the seven candidate peptides

Rank	Peptide sequence	Calculated hydrophobicity	Calculated retention time, min	Observed retention time, min	Integrated abundance
1	SFLVGSAAQSLSK	28.28	20.7	20.3	1,265,000
2	DVTLGSVLGR	24.48	19.0	21.5	495,700
3	IYDQVQSGK	12.36	13.7	13.1	517,300
4	TQEFILNSPTVTSIK	31.91	22.3	22.0	253,000
5	EVAWNLTSVDLVR	37.75	24.9	24.8	220,100
6	GLETTATYDPK	17.12	15.8	18.7	195,100
7	YDGNVYENLFEWAK	44.16	27.8	26.9	30,960

(3) Test the candidate peptides experimentally to determine the best performing peptides and their retention times

Table 3.3 lists the initial set of seven candidate peptide chosen for experimental testing. Of the 30 peptides listed in Table 3.2, 18 are eliminated based on the criteria described above (peptides number 1, 2, 3, 4, 5, 7, 8, 9, 11, 16, 17, 19, 23, 24, 25, 26, 28, 29). Further, an additional four peptides were considered less useful based on containing either cysteine residues (number 13 and 20), a histidine near the C-terminus (number 18), and high molecular weight (number 30). A final peptide (number 21) was eliminated at this stage because of the observation of a related missed cleavage peptide (number 23).

A representative CID spectrum can be obtained for each peptide as a guide to selecting the fragmentation reactions to be used for each. While our laboratory does access and evaluate these CID spectra, we prefer a complete test of all b- and y-ions at this stage. Our initial test includes all y-ions, all b-ions except b_1, and the a_2-ion with default calculated collision energies that are based on the m/z of the parent ions. One could certainly argue the wisdom of not making better use of the reference CID spectrum at this time. However, considering the number of transitions that can be monitored in a single selected reaction monitoring experiment, it simply makes more sense to determine the best reactions experimentally. The key information obtained from the PeptideAtlas reference CID spectra at this time is the observation of any prominent doubly charged ions that should also be included in the initial descriptors. No such doubly charged ions were seen for any of these seven candidate peptides, but they occur frequently enough that the possibility should be considered.

The liquid chromatography retention information is the most critical piece of information that cannot be obtained from the databases. The retention time information is absolutely vital to recognizing the proper chromatographic peak. This recognition is aided by using a test system in which the protein is strongly expressed, but it is our experience that one cannot rely simply on selecting the largest chromatographic peak seen in these initial descriptors.

Table 3.3 also contains the calculated hydrophobicity, calculated retention time, and observed retention time for each candidate peptide. The calculated retention time is based on the hydrophobicity and observed retention time of a series of other peptides seen in samples analyzed by our laboratory. These peptides cover a range of hydrophobicity and are easily seen in common samples. These peptides allow the establishment of a simple linear relationship between calculated hydrophobicity and retention time that is used to calculate the predicted retention time for each new peptide being evaluated. A number of standard peptide mixtures are commercially available for this purpose. Our capillary column LC conditions keep the solvent conditions (composition and gradient) the same at all times but vary the flow rate slightly to target the trypsin autolysis peptide VATVSLPR elution to 17.0 min. Slight variations in the column length lead to a corresponding slight variability in the exact slope and intercept of the hydrophobicity versus retention time line, but the observation of a set of peptides common to many of the samples we analyze makes the relationship straight-forward to establish at any time without any specific effort. In some cases, it may be advisable to exclude a peptide with an observed retention time that is exceptionally different from the calculated time. The rationale for this exclusion is that the aberrant retention time could indicate that an incorrect peptide sequence is being monitored.

Two peptides were eliminated from further consideration at this point based on the data in Table 3.3. The 7th ranked YDGNVYENLFEWAK peptide was eliminated because of low abundance. While optimization could improve the peptide to some extent, the number of far superior peptides makes this elimination a reasonable simplification to make. The 6th ranked GLETTATYDPK peptide was eliminated based on a combination of lower abundance and the larger divergence of the observed retention time from the expected retention time. It is notable that the 2nd ranked DVTLGSVLGR peptide also showed a considerable difference between the calculated and observed retention times. This peptide, however, was significantly more abundant and the nature of the ions observed gave high confidence that the correct peptide was being observed (Fig. 3.4). Therefore, this peptide continues to be considered at this time. However, as seen below, this retention time divergence ultimately makes this peptide less useful compared to other peptides that give similar abundances and have well-behaved retention characteristics. The descriptor development process continued with the five remaining peptides.

(4) Optimize the collision energy for each fragmentation reaction

Table 3.4 summarizes the results of the collision optimization for the peptide IYDQVQSGK. The abundances of the various fragmentation reactions originally observed are shown in the Initial Test Group part of the table.

The collision optimization experiment is based on a separate descriptor generated by the software we use (Pinpoint from ThermoScientific). The optimization experiment begins with the default calculated collision energy and generates an optimization matrix with addition collision energies that vary by 2 eV steps for 6 eV above and below the default value. This gives a total of seven collision energy steps that are tested for each fragmentation reaction for each peptide. The optimum energy is then selected as the energy that gives the largest signal for each reaction. As can be seen, not all reactions have the same optimum collision energy. It is also striking that the

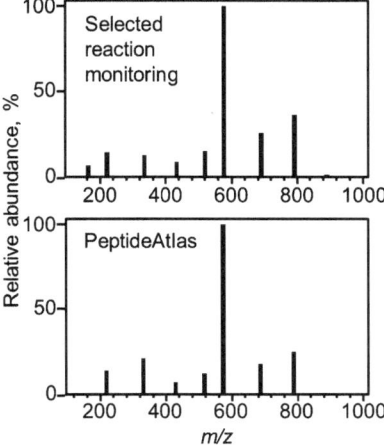

Fig. 3.4 CID spectra of the peptide DVTLGSVLGR. The *top spectrum* is reconstructed from the intensity data acquired by the initial selected reaction monitoring descriptor. The *bottom spectrum* is a reference spectrum obtained from the PeptideAtlas. In each case, only the y-ion series is shown. The similarity of the spectra strongly support detection of the correct peptide despite the late elution versus the calculated hydrophobicity. Not that the y_1-ion at 147 is not detected in the PeptideAtlas spectrum because of the scan range of the ion trap instrument used for the experiment

optimized collision energy produces a significant increase in signal strength for many of the fragmentation reactions. While the increases produced by the optimization may not always be as impressive as seen in this specific example, our overall experience is that collision energy optimization is worth the effort and will give a significant increase in signal strength relative to default conditions.

(5) Select the best fragmentation reactions for the best peptides for the final descriptor

The pattern of abundances seen in Table 3.4 has a couple of characteristics that are both consistent with our general observations for a larger number of peptides and expected based on the gas phase ion chemistry of the collision induced fragmentation of doubly charged tryptic peptides. First, the majority of the more abundant ions will be the y-ion series. These ions are consistently observed and represent the most reliable source of amino acid sequence information in the CID spectra. The second feature, however, is the high abundance of the a_2/b_2-ion pair. The abundance of these ions reflects (a) the equal generation of both a b-ion and a y-ion by each fragmentation reaction for a doubly charged peptide and (b) the ability of larger b-ions to continue to fragment to smaller b-ions when re-activated by multiple collisions, making the a_2/b_2-ions a common end point from the entire b-ion series.

These data are sorted and ranked to choose the reactions shown in the Final Optimized Group half of the table. In this case, we have selected the most abundant fragment ions to include in the optimization. The selection process is helped by

Table 3.4 Ions used in the initial testing of the peptide IYDQVQSGK

Initial test group					Final optimized group			
Fragment ion, m/z	Ion type	Collision energy, eV	Integrated abundance	Rank	Fragment ion, m/z	Ion type	Collision energy, eV	Integrated abundance
147.11	y1	21	38,640	6	147.11	y1	26	56,290
204.13	y2	21	15,190		204.13	y2	26	25,230
291.17	y3	21	51,630	5	291.17	y3	20	60,190
419.22	y4	21	10,910		419.22	y4	18	30,560
518.29	y5	21	16,420		518.29	y5	18	34,070
646.35	y6	21	11,330		646.35	y6	16	35,140
761.38	y7	21	116,500	1	761.38	y7	14	310,300
924.44	y8	21	44,220	4	924.44	y8	18	95,870
249.11	a2	21	167,500	2	249.11	a2	20	290,700
277.15	b2	21	35,710	3	277.15	b2	16	108,500
392.18	b3	21	6,612		392.18	b3	16	18,790
520.24	b4	21	204		520.24	b4	14	6,783
619.31	b5	21	308		619.31	b5	18	4,764
747.37	b6	21	1,093		747.37	b6	14	7,893
834.40	b7	21	848		834.40	b7	14	1,776
891.42	b8	21	185		891.42	b8	14	1,509

The parent ion m/z is 519.27

ranking the abundance of the fragment ions and choosing a consistent cut-off point for further consideration. In this case, the cut-off point is based on selecting the fragment ions that account for 85 % of the total ion current observed monitoring the entire b- and y-ions series. It is interesting to note the differences between this choice versus including a couple additional fragment ions. Specifically, limiting our choice to the top 6, 7, or 8 fragment ions is a difference of accounting for 85 %, 88 %, and 91 %, respectively, of the total ion current. These small incremental steps give a sense of the low relative abundance of the lower ranked fragment ions. The goal is to design a descriptor that collects as much of the total ion current as possible, but it is clear that a point of diminishing return is seen where including more fragment ions does not give a corresponding increase in signal intensity.

Not monitoring ions that do not add significantly to the signal observed for a peptide has two advantages. First, careful selection reduces the number of transitions that are monitored for a peptide. Time spent monitoring unnecessary fragmentation reactions will reduce the overall number of reactions that can be monitored and can increase the potential to detect interfering species. Although no rule exists that states a specific maximum number of reactions, limits are seen based on the need to accurately reconstruct the chromatographic peak and the minimum dwell time of the instrument. Second, limiting the reactions monitored to those with the best contribution to the signal can also lessen the chance of detecting interfering peptides.

The last challenge of the final selection step is evaluating the value of the low m/z ions. This selection is a challenge because, as noted above, the a_2/b_2-ion pair is often high abundance yet the low m/z ions have a considerably greater propensity for

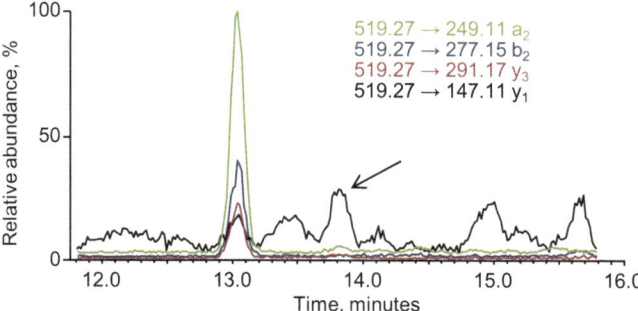

Fig. 3.5 Ion chromatograms for selected low m/z ions in the ACOX1 descriptor. For clarity, only the y_1, y_3, a_2, and b_2 ions are shown. The *arrow* designates the ion chromatogram for the y_1-ion at m/z 147.11

interference and background. Figure 3.5 shows the low m/z ions from the ACOX1 descriptor used in the analysis of a mouse liver sample. The a_2- and b_2-ions at m/z 249.11 and 277.15 (the green and blue traces in the figure) give clear, well-resolved chromatographic peaks with high abundance. No unwanted effects of their inclusion are seen. Similarly, the y_3-ion at 291.17 (red trace) also gives a clear and well-resolved chromatographic peak with unwanted effects. However, the y_1-ion at m/z 147.11 (black trace designated with an arrow) has a similar intensity to the y_3-ion but also has a high relative background seen in the offset from zero and detects a number of additional chromatographic peaks. The high background adds noise to the overall signal. Although these additional peaks are resolved chromatographically, their presence can complicate the subsequent data procession steps such as the automated recognition of the correct chromatographic peak. These issues, combined with the relatively low contribution of the m/z 147 ion to the overall signal, lead us to remove the y_1-ion from the final descriptor.

The final descriptor for ACOX1 is shown in Table 3.5. In this case, the descriptor includes a total of 19 reactions. A key question at this point is, 'Why include this number of transitions for one protein?' Our laboratory prefers to use three peptides for each protein. One important advantage of monitoring multiple peptides is the redundancy of any observations, since the change seen with one peptide should be seen in all peptides. There is also some benefit in being able to exclude a peptide if a problem such as an interfering signal is encountered. For these reasons, three peptides is our preference and we consider two peptides to be a minimum number. We also equate monitoring more peptides to a recording more information about the protein that is measured. The next element is the number of transitions for each peptide. The cut-off used in this example was the transitions needed to record at least 85 % of the total b- and y-ion current. We have tried to standardize this cut-off in order to take advantage of the use of the best flyer concept to estimate an absolute amount of each protein in our sample. This choice also reflects our desire to collect as much information for each peptide as is practical. It is also notable that we used a 3-minute window in the scheduling.

Table 3.5 Final ACOX1 descriptor

	Parent ion, m/z	Fragment ion, m/z	Collision energy, eV	Start, min	End, min	Ion type	Peptide sequence
1	519.27	249.10	20	11.6	14.6	a_2	IYDQVQSGK
2	519.27	277.15	16	11.6	14.6	b_2	IYDQVQSGK
3	519.27	291.17	20	11.6	14.6	y_3	IYDQVQSGK
4	519.27	518.29	18	11.6	14.6	y_5	IYDQVQSGK
5	519.27	761.38	14	11.6	14.6	y_7	IYDQVQSGK
6	647.85	207.10	31	18.8	21.8	a_2	SFLVGSAAQSLSK
7	647.85	235.11	29	18.8	21.8	b_2	SFLVGSAAQSLSK
8	647.85	348.19	21	18.8	21.8	b_3	SFLVGSAAQSLSK
9	647.85	633.36	19	18.8	21.8	y_6	SFLVGSAAQSLSK
10	647.85	791.43	19	18.8	21.8	y_8	SFLVGSAAQSLSK
11	647.85	848.45	19	18.8	21.8	y_9	SFLVGSAAQSLSK
12	647.85	947.52	18	18.8	21.8	y_{10}	SFLVGSAAQSLSK
13	839.46	230.11	35	20.5	23.5	b_2	TQEFILNSPTVTSIK
14	839.46	359.16	35	20.5	23.5	b_3	TQEFILNSPTVTSIK
15	839.46	448.28	31	20.5	23.5	y_4	TQEFILNSPTVTSIK
16	839.46	745.44	27	20.5	23.5	y_7	TQEFILNSPTVTSIK
17	839.46	832.48	25	20.5	23.5	y_8	TQEFILNSPTVTSIK
18	839.46	946.52	27	20.5	23.5	y_9	TQEFILNSPTVTSIK
19	839.46	1,059.60	27	20.5	23.5	y_{10}	TQEFILNSPTVTSIK

There are 19 reactions for this descriptor

All three parameters—number of peptides per protein, number of transitions per peptide, and the time window used to monitor each peptide—directly impact the size of the larger assay that is built. As discussed in Chap. 5, the ACOX1 descriptor is a part of a larger assay used to monitor the expression of the β-oxidation enzymes. The critical limiting factor in the final size of the assay is the need to accurately reconstruct the shape of the chromatographic peak for each peptide. Although no specific rule exists, many investigators would like at least 20 data points across the peak for the accurate and reproducible integration. Considering the time width of the chromatographic peaks of approximately 20 s, this goal means the mass spectrometer must cycle through all transitions being monitored at that time in approximately 1 s. Assuring this cycle time requires management of the number of transitions being monitored at any time using a combination of the number of transitions, the time windows for their monitoring, and the resulting overlap.

The preference in our laboratory is to attempt to manage this cycle time through a combination of the time windows used to monitor each peptide and the number of proteins targeted in the assay. Narrower time windows will reduce the number of transitions being monitored at any point in time. The practical limit is the reproducibility of the retention times, both within a run and between multiple runs. The in-run reproducibility of retention time is generally very good and would allow windows as small as 1 min. The between-run reproducibility, however, is not as good, particularly if one considers them over weeks of time. Wider windows can help minimize the maintenance of the chromatography conditions and descriptor.

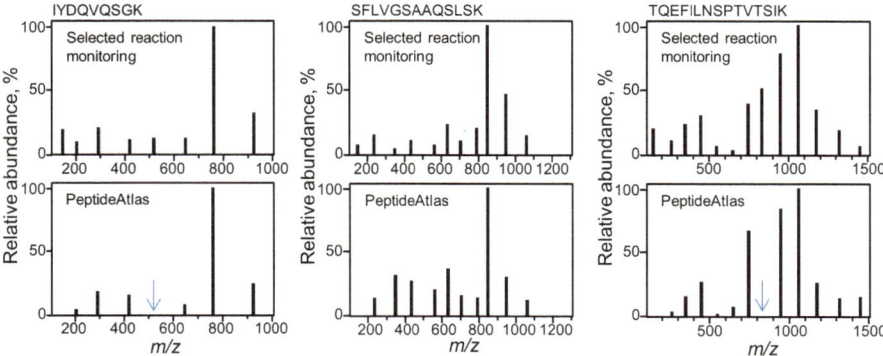

Fig. 3.6 A comparison of CID spectra as a part of the verification process. The *top spectrum* for each is reconstructed based on the abundance of the y-ion series recorded in a selected reaction monitoring experiment. The *bottom spectrum* for each is a representative reference spectrum containing only the y-ions downloaded from the PeptideAtlas. Since these spectra were recorded with an ion trap instrument, the *blue arrow* designate fragment ions that are lost due to the excitation pulse. Further, the low mass cutoff for each left out the y_1-ions at m/z 147

This maintenance includes making fine adjustments to the flow rate to put all peptides in their windows or adjusting the window for all peptides to correct for any changes in retention time. The total number of protein is our β-oxidation assay is 22 and includes the target β-oxidation enzymes, several other related fatty acid processing proteins, two housekeeping proteins, and a non-endogenous internal standard (bovine serum albumin). This set of proteins is detected with a total of 408 transitions managed in a way that gives at least 20 data points across each chromatographic peak.

(6) Validate and test the final descriptor

The final step in the development process involves asking two key questions about how the descriptor is working. Does the descriptor detect the peptides it is designed to detect? How does the descriptor function in the samples that will be analyzed?

The first question is a validation question and is particularly important when the design process begins with information obtained from a database like the PeptideAtlas. Careful skepticism is needed to evaluate the detection of these peptides and specifically the retention time, the CID spectrum, and the apparent abundance of each peptide. Throughout the design process, peptides that elute at times that are consistent with their calculated hydrophobicity are preferred and, in the case of ACOX1 one peptide was excluded because of this issue. It is also possible to compare pseudo CID spectra, reconstructed from the selected reaction monitoring data, with reference spectra obtained from the database. This process is shown in Fig. 3.6. For each peptide, the comparison focuses on the y-ion series because of differences in the b-ion series expected in a triple quadrupole instrument relative to the ion trap and orbitrap instrument that produce most of the data in the database. In all three peptides, strikingly similar spectra are seen even with the lack of the low m/z y_1-ions and the ions lost to the excitation pulse in the ion trap instrument. As a final stage, the BioGPS data are consistent with ACOX1 being significantly more abundant in the liver than in the heart or skeletal muscle and these differences are seen in the LC-tandem mass spectrometry analysis. Additional analyses of sample types that would not contain the specific mouse ACOX1 peptides would also be useful. These samples could come from other mammalian systems or more distant systems such as yeast if available.

As a final validation step, it may be possible to acquire bona fide CID spectra from the final peptides on either the triple quadrupole instrument used for the selected reaction monitoring experiment or another instrument such as an ion trap or orbitrap instrument. These experiments are ideal because they allow the correlation of both a proper CID spectrum and the chromatographic retention pattern in an investigator's own laboratory.

3.5   A Summary of the Descriptor Design Process

Designing the descriptor for each peptide used in a targeted quantitative proteomics assay is the core development activity. It makes sense to use a methodical approach that utilizes all available information, considers all possible peptide fragmentation reactions, and optimizes the best of those reactions. A methodical approach to the development process will have the highest likelihood of finding the peptides that give the best response—the so-called best flyers. It may be entirely possible to streamline the process, but consider a properly designed assay will have long lifetimes after development. It is also a good idea to document the development process in order to facilitate any changes that might be made in an assay. The most common reason for needing some type of change has been the recognition of some undesirable feature as the assay is used over time and with a wider variety of samples. A consistent approach and good documentation will make these changes as smooth as possible.

References

1. Wysocki VH, Tsprailis G, Smith LL, Breci LA (2000) Mobile and localized protons: a framework for understanding peptide dissociation. J Mass Spectrom 35:1399–1406
2. Gucinski AC, Dodds ED, Li W, Wysocki VH (2010) Understanding and exploiting peptide fragment ion intensities using experimental and informatic approaches. Methods Mol Biol 604:73–94
3. Harrison AG, Csizmadia IG, Tang TH (2000) Structure and fragmentation of b2 ions in peptide mass spectra. J Am Soc Mass Spectrom 11:427–436
4. Harrison AG (2009) To b or not to b: the ongoing saga of peptide b ions. Mass Spectrom Rev 28:640–654
5. Lau KW, Hart SR, Lynch JA, Wong SC, Hubbard SJ, Gaskell SJ (2009) Observations on the detection of b- and y-type ions in the collisionally activated decomposition spectra of protonated peptides. Rapid Commun Mass Spectrom 23:1508–14

6. Kinter M, Sherman NE (2000) Protein sequencing and identification using tandem mass spectrometry. Wiley, New York

7. Wu C, Orozco C, Boyer J, Leglise M, Goodale J, Batalov S, Hodge CL, Haase J, Janes J, Huss JW 3rd, Su AI (2009) BioGPS: an extensible and customizable portal for querying and organizing gene annotation resources. Genome Biol 10:R130

8. Desiere F, Deutsch EW, King NL, Nesvizhskii AI, Mallick P, Eng J, Chen S, Eddes J, Loevenich SN, Aebersold R (2006) The PeptideAtlas project. Nucleic Acids Res 34:D655–D658

9. Sievers F, Wilm A, Dineen DG, Gibson TJ, Karplus K, Li W, Lopez R, McWilliam H, Remmert M, Söding J, Thompson JD, Higgins D (2011) Fast, scalable generation of high-quality protein multiple sequence alignments using Clustal Omega. Mol Syst Biol 7:539

10. Spicer V, Yamchuk A, Cortens J, Sousa S, Ens W, Standing KG, Wilkins JA, Krokhin OV (2007) Sequence-specific retention calculator. A family of peptide retention time prediction algorithms in reversed-phase HPLC: applicability to various chromatographic conditions and columns. Anal Chem 79:8762–8768

11. Ludwig C, Claassen M, Schmidt A, Aebersold R (2012) Estimation of absolute protein quantities of unlabeled samples by selected reaction monitoring mass spectrometry. Mol Cell Proteomics 11:M111.013987

Chapter 4
Sample Processing

Abstract The sample processing methods are the critical link between the intact proteins taken from a complex biological system and the mixture of isolated and largely purified peptides injected into the LC-tandem mass spectrometry system. As seen with essentially any advanced analytical technique, the overall success of the selected reaction monitoring experiment is dependent on the quality of these samples. The accuracy and precision of the assay requires that the peptides be generated reproducibly with high yield. The protease trypsin is ideal for this purpose and produces a set of peptides that are uniquely suited to the selected reaction monitoring experiment. The ability to analyze large numbers of samples without unusual effects on the mass spectrometry system requires that confounding classes of biomolecules like lipids, RNA, and DNA be removed. Our laboratory has found the use of acetone precipitation and short-run gel electrophoresis to be useful tools for producing these types of samples.

4.1 Overview

The fundamental result of the sample processing method in a targeted quantitative proteomics experiment is the digestion of the proteins in each sample into a reproducible set of peptides with high yield. Although there are many parallels in the different technical components, the context for the sample processing in a targeted quantitative experiment is markedly different from that of the standard protein identification experiment. The protein identification experiment that has been the basis for proteomics for many years produces a set of qualitative results by searching individual CID spectra (whether 10 or 10,000) against the databases to give a list of component protein identities. These identities may be based on as few as one or two CID spectra, depending of the identification criteria being used and the nature of those CID spectra. Subsequent experiments are carried out to verify those results. These verification experiments will use orthogonal approaches such as Western blot

M. Kinter and C.S. Kinter, *Application of Selected Reaction Monitoring to Highly Multiplexed Targeted Quantitative Proteomics*, SpringerBriefs in Systems Biology, DOI 10.1007/978-1-4614-8666-4_4, © The Authors 2013

and/or molecular biology and may or may not involve additional LC-tandem mass spectrometry experiments. In fact, in most protein identification experiments the combination of multiple high quality CID spectra with clear matches to unique peptides and good sequence coverage will give such a conclusive result that the proteomics experiment is not repeated and all future work is designed to address the biological significance of that result. The practical effect of the isolated nature of the identification experiment is that certain performance characteristics of the LC-tandem mass spectrometry experiment, such as good chromatographic peak shape, reproducible retention times, etc. are often only a modest concern. The point is not that qualitative analyses can be sloppy, but rather to point out that the sample processing associated with these experiments can be streamlined to emphasize factors such as speed, with little effort made to optimize conditions for different sample types.

In contrast, the very essence of quantitative experiments is a comparison between samples with a need to detect and document differences. As a result, quantitative analyses will always involve multiple samples from the different conditions that will be compared plus replicates within those conditions to allow statistical tests. This combination makes the smallest quantitative experiment at least six samples, based on two conditions (for instance control versus treated) with three biological replicates of each for a simple Student's t-test of statistical significance. Far larger experiments with multiple comparisons and greater numbers of biological replicates are common. In all cases, reproducibility in the handling and analysis will be a key part of the overall success of the experiment. Ideally, this handling and analysis will give analytical variability that is small compared to the biological variability to maximize the opportunity to recognize differences in protein expression between the conditions. Successfully achieving this reproducibility requires care, expert technique, and an attention to detail that is a step above that needed for completely successful qualitative experiments.

For this discussion, the sample processing is divided into three stages shown in Fig. 4.1—(a) pre-digestion processing, (b) the protease digestion, and (c) post-digestion processing. The goals of the sample processing are to eliminate compounds that affect the protein digestion, eliminate compounds that will affect the LC-tandem mass spectrometry analysis, concentrate the proteins for digestion, and produce a final sample that will help keep the LC-tandem mass spectrometry experiment as robust and reliable as possible.

4.2 The Pre-Digestion Stage

The challenge and opportunity of sample preparation

The variety of samples encountered in biomedical research creates a great challenge in describing all-encompassing methods, so some context is needed to understand specific approaches. Our laboratory, for example, has processed samples from a number of animal and human tissues, different types of cultured cells, a small number of whole

Fig. 4.1 An overview of the sample processing for selected reaction monitoring. The goals of this process are isolate the proteins form all other components of the sample, digest the proteins to produce a mix of peptides, and transfer those peptides from the digest into a sample that is suitable for LC-tandem mass spectrometric analysis

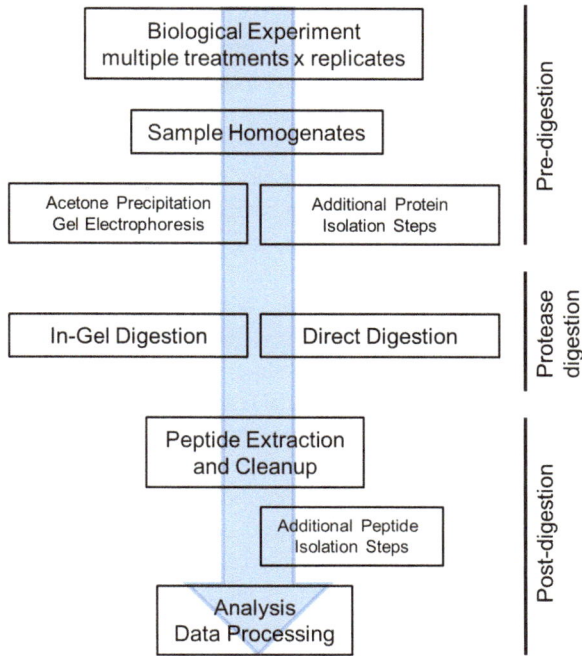

organisms such as worms (C. elegans) and fish (D. rerio), and various laboratory preparations such as immunoprecipitates. Most of our work, however, has been with animal tissues and the following discussion will focus on these methods with an emphasis on mouse heart and skeletal muscle tissues. As will be seen, a key point in the processing of these samples that we believe extends quite well to all types of protein sample is a protein precipitate made by treating the sample using acetone.

A first step in sample processing is producing a representative homogenate. Often the homogenization step will seek the complete dissolution of the tissue in a detergent containing buffer. The most commonly used detergent is arguably sodium dodecyl sulfate (SDS), but other detergent mixtures like RIPA buffer that contains a mixture of SDS, sodium deoxycholate, and Triton X-100 are also used effectively. In other cases, the tissue may be homogenized in buffer only, without detergent. In all cases, the homogenization process is often aided by mechanical devices like a Teflon Potter-Elvehjem homgenizer. It may also be advantageous to heat the samples, although care should be taken that any heating is compatible with other uses of the sample and all reagents.

Two general homogenization protocols

A key question in designing the homogenization protocol is whether the samples that are produced will also be used for other types of analyses, such as enzymatic assays, sub-cellular fractionation, or even RNA isolation. Often these additional uses will be the deciding factor in how the sample is handled. This is the case in the mouse heart homogenization procedure used in our laboratory. Since (a) mice have only one heart, (b) the heart has specialized anatomical parts that are not equivalent, and (c) we sought to produce parallel results between the proteomics and functional assays, all of our mouse heart experiments have used this homogenization procedure for the initial handling of mouse heart tissue. In this procedure, the hearts are homogenized without detergent in buffer containing 10 mM MOPS, 1 mM EDTA, 210 mM mannitol, 70 mM sucrose, pH 7.4, with a Teflon Potter-Elvehjem homogenizer. The crude homogenate is cleared by centrifugation at $550 \times g$ for 5 min, discarding the pellet. We refer to this initial homogenate as the S1 (first supernatant) homogenate. Mitochondria are then isolated from the S1 homogenate by centrifugation at $10,000 \times g$ for 10 min to pellet. The supernatant is discarded and the pelleted mitochondria resuspended in the MOPS/EDTA/mannitol/sucrose buffer described above. Both the S1 homogenate and the suspended mitochondrial remain biochemically functional and are used for a variety on enzymatic and mitochondrial function assays.

For the quantitative proteomics experiment, volumes of the S1 homogenate or the mitochondrial preparation containing 60 µg protein (typically approximately 100 µL volumes) are mixed with 10 % SDS in water to give a final SDS concentration of 1 %. An amount of the non-endogenous internal standard bovine serum albumin (BSA) is added. We add 8 pmol of BSA based on adding 20 µL of a 25 µg/mL solution. The sample is then heated in a 70 °C water bath for 15 min to assure complete dissolution of the proteins and equilibration of the internal standard. A couple of key points can be made about these initial steps. First, consistent and accurate protein assays should make taking the 60 µg sample a very accurate starting point. Tissues give generous amounts of protein and the protein assay methods are well-tested. Second, the bovine serum albumin standard is easily standardized by careful weighing of the inexpensive and plentiful protein, allowing accurate additions of a known amount. Third, the combination of the aqueous detergent mixture and heating allows equilibration between the internal standard and the sample proteins. Clearly, other types of samples will require modification to this approach, but critical goals of any method should be accuracy of the amount of protein taken for analysis, accuracy in the amount of internal standard added, and the opportunity for effective equilibration of the two.

In the case of skeletal muscle tissues (as an example), each mouse has multiple muscles that are anatomically identical. For example, each leg has a gastrocnemius muscle that represents essentially identical biological samples. It is, therefore, easy to take one muscle for targeted quantitative proteomics and the other muscle for the mitochondrial isolation and functional assays. The muscles are approximately 100 mg wet weight, are diced and mixed with 5 mL of a reagent containing 1 % SDS, 1 mM EDTA, 50 mM Tris pH 7.8. The mixture is homogenized with a Teflon Potter-Elvehjem homogenizer and heated in boiling water for 15 min. The sample is cleared by centrifugation at $550 \times g$, aliquots of the supernatant collected, and a protein assay used to determine the protein concentration.

In some experiments it may be desirable to purify subgroups of proteins. The isolation of mitochondrial proteins describe above is one such example.

Fig. 4.2 Short-run gel electrophoresis. Representative gels for (**a**) heart, (**b**) liver, (**c**) skeletal muscle. These gels were stained for 1 h to achieve a consistent result. For samples used for selected reaction monitoring, minimal staining times of approximately 5 min are used to help reduce the subsequent destaining times

These subcellular fractionation methods required intact subcellular organelles which limits how the homogenization can be done. Other techniques, however, can still use sample that have been processed via the stronger denaturing final stages of the homogenate preparation.

Acetone precipitation and short-run gel electrophoresis

The final steps in the pre-digestion processing used by our laboratory are an acetone precipitation and short-run gel electrophoresis. Acetone precipitation of small amounts of protein is an effective concentrating and desalting procedure. For the precipitation, a volume of ice-cold acetone is added to give a final concentration of >80 % acetone. We typically add 1 mL acetone to our 100–200 μL samples. Although the precipitation is likely complete in a couple of hours, our protocol allows the precipitation to go overnight at −20 °C. The proteins are pelleted by centrifugation at $10,000 \times g$ for 10 min and the acetone gently poured off. Any residual acetone is removed in a SpeedVac. This combination of gentle pouring and final evaporation means that the small, essentially invisible protein pellet is not disturbed yet at least 90 % of the acetone-soluble material is removed. In cases where we then digest the protein with trypsin without the intervening short run gel electrophoresis described below, multiple rounds of acetone washing and pelleting are used to complete the SDS removal.

Short-run gel electrophoresis is an otherwise standard SDS-PAGE experiment, with the key modification that the samples are only run a minimal distance into the gel. This variation was first described by Mann for the effective in-gel digestion of the major pre-mRNA spliceosome [1]. For this procedure, the precipitated protein is dissolved in Laemmli sample buffers at protein concentration of 1 μg/μL and a 20 μL aliquot run 1.5 cm into the resolving component of a 12.5 % SDS-PAGE gel (Fig. 4.2). The gel is subsequently fixed and stained with Coomassie blue according to standard procedures. As can be seen in the figure, some separation of the proteins is observed but this separation is largely irrelevant. Two more important characteristics are the ability to see equal staining of all samples and the immobilization of the proteins in each sample in a small section of gel. The equal staining is expected based on the equal loading of 20 μg protein in each lane. The actual observation of equal staining, however, is a key quality control point in the overall workflow of the analysis. In fact, this ability to directly observe the integrity of each sample is an important result of the short-run gel electrophoresis step. The immobilization of the proteins in a small amount of gel is important because it allows a small scale in-gel digestion procedure to be used. The details of the in-gel digestion are described below but the salient feature here is the ability to cut the entire lane as a single sample and easily wash all sample preparation reagents out to the sample. These reagents include any SDS, other gel electrophoresis reagents, any sample preparation reagents still in the sample after protein precipitation and the Coomassie blue stain. Ultimately, the short run gel electrophoresis presents verifiably equal amounts of denatured protein immobilized in an inert plastic matrix for the digestion procedure and, ultimately, the LC-tandem mass spectrometric analysis.

4.3 Protease Digestion

Trypsin is an ideal proteolytic enzyme for proteomics

Essentially all proteomics experiments, including targeted quantitative proteomics, use the enzyme trypsin to digest proteins into a mixture of peptides. Some of the advantages of trypsin are practical. Trypsin is easily isolated from porcine liver, making it an inexpensive enzyme to isolate and purify extensively. A number of well-characterized modifications and treatments have been developed that minimize the other proteolytic reactions. The more important advantages, however, relate to the specific amide bonds trypsin hydrolyzes when digesting a protein. The amide bonds that trypsin cleaves are limited to the C-terminal bonds of lysine and arginine, except when the adjacent amino acid is a proline. In one-letter amino acid codes this specificity is expressed as cleaving K–X and R–X bonds except where X = P.

The first important advantage of the specificity of trypsin is that for the many high quality, commercially-available, proteomics-grade preparations, the specificity is essentially absolute. Few competing reactions are observed with even high concentrations of the enzyme and long reaction times. In other words, the tryptic digestion can be driven to completion without any concern for secondary reactions taking place. This is not to say that all expected bonds will be cleaved with 100 % efficiency because they are not. All proteins will contain a subset of unfavorable bonds that are poorly cleaved and seen as *missed cleavages*. The more important fact, however, is that all proteins will also contain a majority of favorable bonds that are cleaved with 100 % efficiency. This preponderance of favorable cleavage sites and the ease of driving the trypsin digestion to completion, gives a high degree of reproducibility to the set of peptides that are produced in the digestion of a protein. These peptides are seen in the coverage of the protein sequence and are the actual targets in the targeted quantitative proteomics experiment.

The second important advantage of the specificity of trypsin is the generalized amino acid sequence that tryptic peptides will have. This sequence features one basic site at the N-terminus of the peptide and a second basis site on the side chain of the lysine or arginine residue at the C-terminus of the peptide. Under electrospray ionization, these peptides will favor doubly charged molecular ions with one proton localized at the N-terminus and one proton localized at the C-terminus [2]. Importantly, this charge structure drives an informative set of fragmentation reactions that favor efficient production of complementary y- and b-ions with low-energy collision conditions. In addition, codon usage shows that approximately 1/9th of the amino acid content of mammalian proteins with be either lysine or arginine residues meaning that the average length of a tryptic peptide will be a 9-mer and have a molecular weight in the range of 1,200 Da. This combination of peptide molecular weight and doubly charged ions fragmenting in a predictable manner is the basis of the experimental design process describe in Chap. 3.

A general in-gel trypsin digestion protocol

The sample preparation protocol described above results in a series of protein samples immobilized by fixation and stained with Coomassie blue in a piece of polyacrylamide gel with approximate dimensions of 10 mm high × 8 mm wide × 1 mm thick. The dimensions are equal to a volume of 80 mm^3 or 80 µL. The protocols described put 20 µg of total protein in the gel. The Coomassie staining shown in Fig. 4.2 allows one to accurately cut the entire protein sample from the gel as a single gel piece using a scalpel. This gel piece is then divided simply into approximately 10 smaller pieces that are placed in a pre-rinsed 1.5 mL microcentrifuge tube. There is no need to divide the gel very finely, mash, or otherwise pulverize the gel. Not only is there no gain from these steps but they increase the opportunity to lose gel pieces during the subsequent handling steps and bias the results. A recurring part of the procedure is the need to add reagents to the samples and then remove them from the tube. These steps are facilitated by the ability to see the get pieces during the handling.

The first part of the procedure is to wash away the Coomassie blue. This washing/destaining step is often the most time consuming part of the process. In fact, we have shortened our staining time to approximately 5 min to give the minimal amount of staining needed to document the sample and guide the cutting process. The washing/destaining uses 1 mL amounts of 50 % ethanol, 40 % water, 10 % acetic acid. The process is accelerated by heating in a 50 °C water bath overnight and several changes of the reagent the following day are used to completely remove the stain. We always make the effort needed to completely destain the sample, believing that this effort gives a sample in which all remnants of the sample processing and gel electrophoresis steps have been completely removed. The sample is then dried in a SpeedVac to remove any residual acid.

The proteins are then reduced with dithiothreitol and alkylated with iodoacetamide. For the reduction step, 200 μL of 10 mM dithiothreitol in 50 mM ammonium bicarbonate is added for 15 min at room temperature. This reagent is removed and the sample is alkylated in 200 μL of 50 mM iodoacetamide in 50 mM ammonium bicarbonate at room temperature for 15 min. These reagents are washed away with successive 5 min treatments with 500 μL ethanol, 500 μL 50 mM ammonium bicarbonate, and 500 μL ethanol. The gel samples are dried in the SpeedVac to prepare for the trypsin incorporation.

The trypsin solution is 20 μg proteomics grade trypsin in 4 mL 50 mM ammonium bicarbonate. We add 200 μL amounts, 1 μg total trypsin to each dried sample for the digestion. The dried gel pieces adsorb the solution to bring the trypsin into the matrix of the polyacrylamide gel. The samples are incubated overnight at room temperature to complete the digestion. The following day, the peptides are extracted from the sample in 2, 200 μL aliquots of an extraction reagent (70 % methanol, 25 % water, 5 % formic acid). These extracts are combined in a rinsed 1.5 mL microcentrifuge tube, evaporated to dryness in a SpeedVac, and reconstituted in 100–200 μL of 1 % acetic acid for the LC-tandem mass spectrometric analysis.

A couple of notes about this procedure, (a) a key aim is consistency in all aspects of the overall process, and (b) some investigators use combination of trypsin and Lys C because Lys C will cut K–P bonds. Consistency in the process includes both quantitative points (the use of 20 μg total protein, the addition of 8 pmol bovine serum albumin as an internal standard, the use of 1 μg trypsin for the digestion) and qualitative points (complete removal of the Coomassie blue, all reagents are made fresh). Our observation that this consistency enhances all aspects of the overall experiment. As discussed in Chap. 5, sample and analysis quality are easily monitored by consistent signals for the trypsin autolysis and the bovine serum albumin peptides. Further, the emphasis on sample washing enhances chromatographic column lifetimes and minimizes source contamination to produce more consistent instrument performance over longer periods of time. We do not use Lys C. Although we have not specifically tested its use, our experience is that most proteins produce a sufficient number of amenable peptides with trypsin alone that it is hard to expect much added value for the combined digestion method.

Direct digestion as an alternative approach

It is our experience that the acetone precipitation and short run gel electrophoresis combine to give a fundamental similarity to a variety of samples from a variety of sources. This similarity greatly facilitates all other parts of the experiment from the digestion procedure to the LC-tandem mass spectrometric analyses to the data-processing steps. It should be noted, however, that digestion without the gel electrophoresis is also used by many investigators. The key accommodations for denaturing, reducing, and alkylating the proteins are needed in these procedures to assure complete and reproducible digestion. These accommodations, however, introduce the new problem of removing those reagents prior to the digestion and LC-tandem mass spectrometry steps. This removal is particularly important for any detergents or chaotropic agents used denature the proteins. Almost all such reagents will hurt the activity of the proteases. Trypsin can tolerate low concentrations of urea (<1 M) but does not tolerate SDS. Similarly, LC-tandem mass spectrometry can tolerate low concentrations of urea, but does not function for SDS containing samples because of fundamental issues in the electrospray ionization. As a rule, direct digestion and analysis of crude protein samples is highly susceptible to deleterious effects of an uncountable list of reagents including salts, detergents, lipids, DNA, RNA, sugars, and more. Our protocol for direct digestion of protein mixtures uses acetone precipitation and washing to accomplish this important task.

One modest advantage of the direct digestion is that without the gel step, more protein can be processed. While gel loading limits sample sizes to approximately 20 µg total protein, depending the characteristics of the gel. Our direct digestion protocol, in contrast, uses 100 µg total protein. For heart homogenates, we could not use the MOPS/EDTA/mannitol/sucrose buffer described above. Instead, the homogenization was carried out in 10 mM Tris, pH 8.0. While these homogenates can be used for enzyme activity assays, we have not used them for subsequent mitochondria isolation. Volumes equivalent to 100 µg total protein are taken, mixed with 10 % SDS to give a final concentration of 1 % SDS, and the bovine serum albumin internal standard added. The sample is heated in a 70 °C water bath for 10 min to denature the proteins and equilibrate the internal standard. The proteins are reduced by adding 50 µL of 50 µM DTT and reacting for 15 min in a 70 °C water bath. The proteins are alkylated by adding 50 µL of 100 µM iodoacetamide and reaction for 15 min at room temperature. Acetone precipitation is used to concentrate the protein and remove the reagents. The initial precipitation is overnight at −20 °C as described above. However, the dried protein that is produced is washed two times with acetone to remove any residual reagents, but particularly any residual SDS. For each wash, 1 mL of ice-cold acetone is added but not mixed, the sample incubated in at −20 °C for 30 min, and the protein pelleted by centrifugation at $10,000 \times g$ for 10 min. For the trypsin digestion, some extra care is taken to put the protein back into the solution. A small amount of acetonitrile is added (50 µL) and the sample vortexed vigorously. A solution of ammonium bicarbonate (200 µL of 100 mM in water) is added and the sample heated in a 70 °C water bath for 15 min. Finally, the sample is cooled to room temperature before adding 1 µg trypsin in 200 µL of 100 mM ammonium bicarbonate buffer for the protein digestion.

4.4   Post Digestion Processing

Our post digestion processing of the samples is quite simple. Peptides are extracted out the gel pieces in a solution of ethanol/water/formic acid (70/20/10). Other extraction systems are equally effective with the key components being a solvent that will help elute the peptides off any active sites in the gel or polypropylene tube and a small amount of an organic acid to facilitate this process. We favor ethanol because it is a common solvent in our laboratory that is used for multiple purposes.

We have found methanol or acetonitrile equally effective. Formic acid is used because of a modest improvement in evaporation time. Again, we have found acetic acid equally effective. Other laboratories have reported the use of SDS removal beads or a combination of acid and basic pH extraction buffers for various reasons. In our hands, however, these steps do not have any advantages for the samples we generate. In fact, the primary property of the extraction system that has driven our choices is the desire to minimize the time required to evaporate the extract before the final reconstitution in 1 % acetic acid for the LC-tandem mass spectrometry experiment.

4.5   Other Methods

Our use of short-run gel electrophoresis is an unusual step in the preparation of samples for a selected reaction monitoring experiment. Beyond this, however, the methods used by other scientists have the same need to denature the proteins, reduce, alkylate, and digest with trypsin. The key parts of these other methods show some variation in how the denaturing is handled and several innovative approaches to the post digest processing. It is informative to compare and contrast these literature reports as a part of designing one's own laboratory method.

Anderson described the quantitation of 53 plasma proteins [3]. For that method, the plasma was initially treated with an affinity purification system to deplete the sample of high abundance plasma proteins such as albumin and immunoglobulin. The sample was then desalted and exchanged into an ammonium bicarbonate buffer using a spin column concentrator. An SDS-containing tris buffer was added and the samples heat denatured before reduction, alkylation, and digestion with trypsin. No specific reference is made to any attempts to remove the SDS prior to the otherwise standard reversed-phase LC-tandem mass spectrometric analysis. Muddiman has also reported the analysis of a protein biomarker in serum, prostate specific antigen, by selected reaction monitoring [4]. Again, an in-solution digestion was also used with the largest difference being treatment with 6 M urea to denature the proteins. In experiments looking at metabolic pathways in cultured cells, Diamandis used an acid-cleavable, electrospray-compatible detergent for the denaturing step [5]. After reduction, alkylation, and trypsin digestion, the detergent was cleaved with formic acid and the peptides isolated from the mixture using solid phase extraction. Based on a survey of the literature, these approaches appear typical and are consistent with in-solution digestion methods with the primary difference being the reagent used for the key denaturing step.

In contrast to our simple approach to post-digest processing, other laboratories have reported highly innovative methods to enhance the analysis by extracting important subsets of peptides out of the digest. In the so-called SISCAPA experiment (stable isotope standards with capture by anti-peptide antibodies) described by Paulovich, an in-solution digestion with SDS-based denaturing was used [6]. Additional steps were used to affinity isolate targeted peptides and their isotopologues. The digest was desalted by solid-phase extraction on a C18-column. The eluate from the

C18-column was mixed with the respective anti-peptide antibodies in phosphate buffered saline and CHAPS. The antibody bound peptides were pulled from the mixture with Protein G-coated magnetic beads and the peptides eluted in an acetic acid/CHAPS mixture. Similarly, White used a combination of anti-phosphotyrosine antibodies and immobilized metal affinity column (IMAC) enrichment in a selected reaction monitoring analysis of phosphopeptides [7]. These experiments used urea to denature the proteins for digest. The resulting peptides were purified by solid phase extraction. The extracted peptides were applied to an antiphosphotyrosine antibody column, eluted with glycine, and the eluate applied to the IMAC column. Other laboratories have reported using avidin isolation of biotinylated peptides in an isotope coded affinity tag experiment with detection in a selected reaction monitoring experiment [8] and the use of peptide specific aptimers for selected reaction monitoring [9]. A common feature of all of these experiments that used affinity isolation of specific peptides was the addition of stable isotope labeled standards for each peptide to rigorously account for any variation in the recovery.

4.6 Summary

The goals of this chapter were to give specific ideas about how our laboratory processes samples, with at least some insight into the logic we have used during the development, and provide some links to the literature to see how other laboratories operate. It is clear that the various methods share a number of common traits and have several pointed differences. As described, key elements are the need to isolate proteins from other parts of the sample; reduce, alkylate, and digest the protein; and process that digest into an appropriate form for LC-tandem mass spectrometry. Many of the differences are in the details of the reagents that are used. In some ways, the fact that all of these reported methods end in a high quality assay is a tribute to the latitude one has in making these choices. This latitude, in turn, can shift the emphasis to considerations of efficiency and reproducibility based on the resources and sensibilities of individual laboratories.

References

1. Rappsilber J, Ryder U, Lamond AI, Mann M (2002) Large-scale proteomic analysis of the human spliceosome. Genome Res 12:1231–1245
2. Covey TR, Huang EC, Henion JD (1991) Structural characterization of protein tryptic peptides via liquid chromatography/mass spectrometry and collision-induced dissociation of their doubly charged molecular ions. Anal Chem 63:1193–1200
3. Anderson L, Hunter CL (2006) Quantitative mass spectrometric multiple reaction monitoring assays for major plasma proteins. Mol Cell Proteomics 5:573–588
4. Barnidge DR, Goodmanson MK, Klee GG, Muddiman DC (2004) Absolute quantification of the model biomarker prostate-specific antigen in serum by LC-Ms/MS using protein cleavage and isotope dilution mass spectrometry. J Proteome Res 3:644–652

5. Drabovich AP, Pavlou MP, Dimitromanolakis A, Diamandis EP (2012) Quantitative analysis of energy metabolic pathways in MCF-7 breast cancer cells by selected reaction monitoring assay. Mol Cell Proteomics 11:422–434
6. Whiteaker JR, Zhao L, Anderson L, Paulovich AG (2010) An automated and multiplexed method for high throughput peptide immunoaffinity enrichment and multiple reaction monitoring mass spectrometry-based quantification of protein biomarkers. Mol Cell Proteomics 9: 184–196
7. Wolf-Yadlin A, Hautaniemi S, Lauffenburger DA, White FM (2007) Multiple reaction monitoring for robust quantitative proteomic analysis of cellular signaling networks. Proc Natl Acad Sci USA 104:5860–5865
8. Arnott D, Kishiyama A, Luis EA, Ludlum SG, Marsters JC Jr, Stults JT (2002) Selective detection of membrane proteins without antibodies: a mass spectrometric version of the Western blot. Mol Cell Proteomics 1:148–156
9. Zhao Y, Widen SG, Jamaluddin M, Tian B, Wood TG, Edeh CB, Brasier AR (2011) Quantification of activated NF-kappaB/RelA complexes using ssDNA aptamer affinity-stable isotope dilution-selected reaction monitoring-mass spectrometry. Mol Cell Proteomics 10: M111.008771

Chapter 5
Sample Analysis and Data Processing

Abstract The application of a selected reaction monitoring method in a complete, multiplexed assay is achieved by combining multiple descriptors. Care must be taken to find an optimal number of targets so that proper sampling of the chromatographic experiment is still obtained. The final challenge is then analyzing the data to determine the signals for the respective peptides and calculating an amount of each protein. Internal standards and housekeeping proteins facilitate this process.

5.1 A Multiplexed Assay for Proteins in the β-Oxidation of Fatty Acids

The descriptor design process detailed in Chap. 3 focused on the protein ACOX1 as an example. ACOX1 is one enzyme in a series of reactions that process fatty acids for energy. We will continue to focus on this group of proteins to illustrate and discuss the use of several selected reaction monitoring descriptors in a larger targeted quantitative proteomics assay for a group of proteins.

As a reminder, a distinctive feature of the selected reaction monitoring method for quantitative proteomics is the ability to multiplex the analysis. Multiplexing is accomplished by assembling the descriptors for multiple proteins into a single assay. The rationale for including any specific protein in an analysis is as varied as the applications. For example, one key application of targeted quantitative proteomics is to perform a second level validation of a biomarker discovery result [1–3]. In these experiments, the proteins are included in the assay based results from a discovery experiment. This use of selected reaction monitoring as a follow-up for an unbiased discovery experiment represents a clear step into the role currently held by immunochemical methods such as Western blot and ELISA. Others have used selected reaction monitoring for the measurement of proteins with known clinical importance, demonstrating the utility of the method in a clinical laboratory setting [4, 5].

M. Kinter and C.S. Kinter, *Application of Selected Reaction Monitoring to Highly Multiplexed Targeted Quantitative Proteomics*, SpringerBriefs in Systems Biology, DOI 10.1007/978-1-4614-8666-4_5, © The Authors 2013

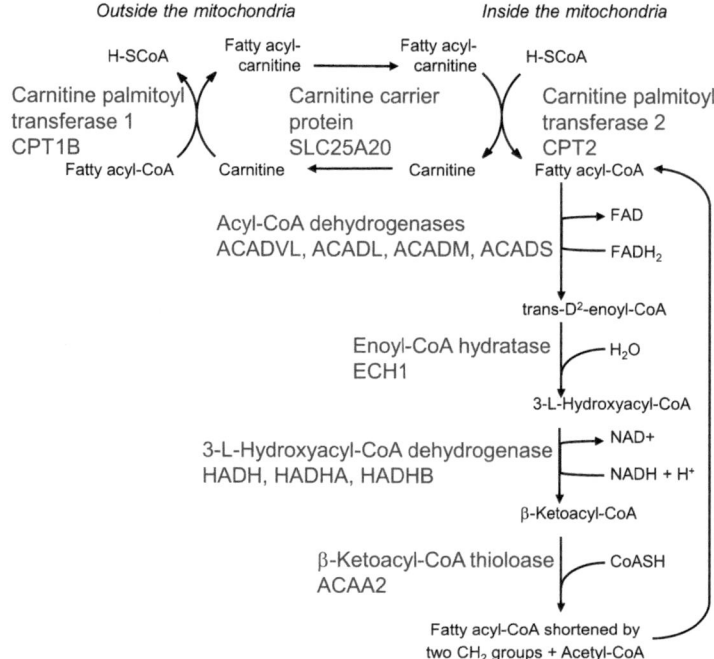

Fig. 5.1 A schematic overview of the β-oxidation pathway. The protein targets in this pathway are shown in *blue* with both the descriptive name and the gene symbol. A total of 12 proteins make up the canonical β-oxidation pathway

In our laboratory, we have been using a different rationale. Our assays are designed based on knowledge of different metabolic pathways that are related to the pathobiology being studied. For example, one set of experiments in our laboratory is evaluating the effects of high fat diets, and the resulting obesity, on the biochemistry of the heart [6, 7]. These experiments have led to a desire to understand how a high fat diet affects the utilization of lipids and carbohydrates in different mouse tissues, but most notably the heart, skeletal muscle, and liver. One result of this work was an observed change in mitochondrial respiration with lipid and carbohydrate substrates. Therefore, assays to monitor changes in expression of the enzymes used for lipid and carbohydrate metabolism, with a focus on mitochondrial function, were designed.

For the lipid metabolism assay the design process began with known components of the β-oxidation of fatty acids shown in Fig. 5.1. These reactions occur in the mitochondria and produce acetyl-CoA which enters the Krebs cycle for complete oxidation that drives ATP production. The goal of this design rationale is to interrogate an entire pathway to produce a quantitative picture of the expression of every protein in that pathway. This approach combines the quantitative analysis of protein expression with elements of discovery since the only prior information used is the knowledge of the pathway components. Other investigators have also taken a similar pathway interrogation approach to assay design [8].

Table 5.1 The proteins that make up the β-oxidation pathway assay

Central β-oxidation enzymes		Other proteins	Housekeeping
ACAA2 (3)	HADHA (3)	ACSL1 (3)	HSPD1 (3)
ACADS (3)	HADHB (3)	ACOT13 (3)	VDAC1 (3)
ACADM (3)	SLC25A20 (2)	CD36 (3)	
ACADL (3)		CRAT (3)	**Internal standard**
ACADVL (3)	**Associated proteins**		BSA (3)
CPT1B (3)	ACOX1 (3)		
CPT2 (3)	DECR (3)		**Extra**
ECH1 (3)	ECI (2)		Trypsin (2)
HADH (2)			

Each protein is listed as its gene name with the number of peptides monitored in parentheses

As shown in Fig. 5.1, the assay design must include 12 proteins to cover the canonical β-oxidation reactions. As will be discussed later, we also add 3 proteins; heat shock 60 kDa protein 1, chaperonin (HSPD1), voltage-dependent anion channel 1 (VDAC1), and bovine serum albumin (BSA) as housekeeping proteins and an internal standard. HSPD1 and VDAC1 are the housekeeping proteins and are a part of all our mouse heart, skeletal muscle, and liver assays. BSA is included as a non-endogenous internal standard and a known amount is added to all samples. This initial group of 15 proteins is measured based on a total of 43 peptides using a total of 282 reactions. Six additional proteins are added based on several interests. Acyl-Coenzyme A oxidase 1, palmitoyl (ACOX1), enoyl-Coenzyme A delta isomerase 1 (ECI), and 2,4-dienoyl CoA reductase 1, mitochondrial (DECR1) are added as other components of β-oxidation. ACOX1 catalyzes the same reaction as the acyl-CoA dehydrogenases, although primarily in peroxisomes. ECI and DECR1 are accessory enzymes for the β-oxidation pathway needed to accommodate the double bonds in unsaturated fatty acids. The last four proteins included in the assay, CD36 antigen (CD36), carnitine acetyltransferase (CRAT), acyl-CoA synthetase long-chain family member 1 (ACSL1), and acyl-CoA thioesterase 13 (ACOT13), are included as proteins with other interesting roles in fatty acid processing.

The complete group of proteins included in the β-oxidation assay is shown in Table 5.1. The quantitation of these proteins is based on 63 peptides and a total of 408 fragmentation reactions.

The number of fragmentation reactions in this assay is at the high end of what we are comfortable monitoring in our system. The limiting factor for determining this number is the cycle time needed to move through all reactions being monitored at one time. This rate, in turn, can be limited by the dwell time of the instrument needed to record the signal precisely and the time needed to step to the next reaction condition. Monitoring too many fragmentation reactions at once will have the undesirable and interrelated effects of reducing dwell times and/or increasing the cycle times. Since dwell time is the time used to record the signal, short dwell times can reduce the signal-to-noise ratio. The more noticeable effect, however, will often be increased cycle times. Figure 5.2 shows the effects of two different cycle times on the reconstruction of a chromatographic peak. Example A reflects the general

Fig. 5.2 Detrimental effects of slow sampling rates on chromatographic peak reconstruction. (A) An exemplary chromatographic peak sampled ideally with 20 data points across the peak. (B) The same chromatographic peak with only 10 data points across the peak. Poorer reconstruction of the peak shape is seen in three places: (1) in the leading edge, (2) at the peak top, and (3) in the trailing edge

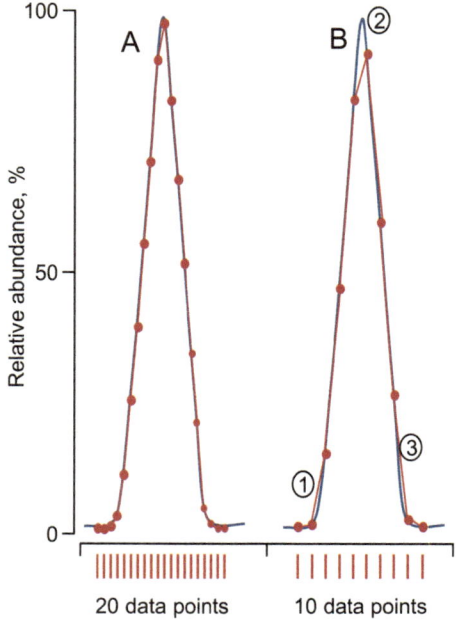

consensus that at least 20 data points are needed to accurately reconstruct the shape of a chromatographic peak from mass spectrometry data. In this example, the connections between the data points closely follow the idealized shape of the peak and the integrated chromatographic peak area under the reconstructed curve will accurately reflect the abundance of that peptide. Example B, on the other hand, shows the effects of having only 10 data points across the same chromatographic peak. Areas of particular concern are the poor reconstruction of the leading edge, the missed peak top, and the poor reconstruction of the trailing edge. Inaccuracy in these areas directly translates into inaccuracy in determining the integrated chromatographic peak area. Further, slight variations in the timing of the chromatographic peak relative to the specific time points that are sampled produces variations in the values and poor precision. It is our experience that the number of points recorded across a chromatographic peak should be actively monitored during the development of an assay since some elements of the cycle time may be outside of operator control. Our instrument, for example, has a minimum dwell time which forces the cycle time to increase despite specifications to the contrary. Other parameters such as number of microscans can be changed unknowing during other tasks such as tuning and calibration.

There are several choices that can be made in the management of cycle time. The first choice is in the time window scheduling for each peptide as a way to minimize the number of reactions being monitored at a time (Fig. 5.3). The narrower these time windows are set, the more peptides can be included in an assay. In a small number of cases, one may also recognize opportunities to select peptides that elute at times that may be distinctive, such as earlier or later than most. Narrower time windows can also have a tangible benefit in the data analysis since other potentially

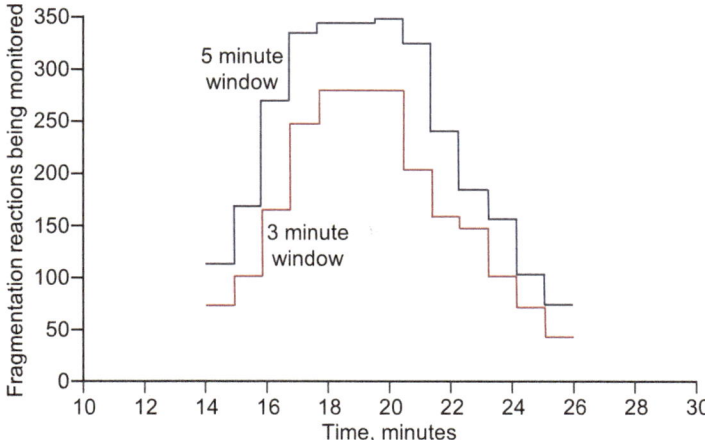

Fig. 5.3 The number of fragmentation reactions being monitored at different times during a chromatographic run. In a highly multiplexed assay, the number of fragmentation reactions being monitored can be managed by altering the time window for each peptide

confounding chromatographic peaks will not be recorded. A limiting factor in the narrowness of the time window, however, is the reproducibility of the retention times, both within and between sets of samples. The within set reproducibility tends to be the easiest to recognize since care must be taken to keep peptide retention times from drifting outside the established time windows. The between set variability, however, can be more complex and reflect both issues with the lab environment like temperature and other factors like new columns or new preparations of the solvents. Overall, the practical limit for the width of the time window will often be how much effort is lost rerunning samples due to missed peptides or maintaining the assay by needed adjustments of the retention times.

The second choice then becomes the combination of deciding how many proteins to include and how many peptides per protein. Our laboratory always keeps the set of fragmentation reactions needed for the 85 % threshold targeted in the descriptor design process. Our preference is to include three peptides per protein, except in cases of small proteins where suitable peptides cannot be found. We have, however, obtained excellent results with methods based on two peptides per protein, making that reduction a reasonable compromise when needed. Although other published assays use only a single peptide per protein, we have not used less than two peptides. As the maximal number of peptides and proteins is reached, a better alternative is to simply move the additional proteins to another assay. As noted in the sample preparation method, digests are reconstituted in 150 μL total volume. Since the liquid chromatography system makes 10 μL injections, this sample volume can accommodate at least 14 total assays. The obvious drawback to this choice is the additional time required to reanalyze the samples. In summary, managing the cycle time of the assay begins with the need to preserve proper reconstruction of the chromatographic peaks and can be accomplished through the optimization of a number of assay components.

5.2 Housekeeping Proteins and Non-Endogenous Internal Standards

A constant part of quantitative analyses that use mass spectrometry is the incorporation of some sort of internal standard into the samples and the analysis. The role of the internal standard is to compensate for all variation during the sample processing and during the final measurement. Internal standards are a part of many quantitative methods, but they are particularly important in mass spectrometry experiments because of the combination of the multi-step sample preparation protocols and the natural day-to-day variability of such a high sensitivity method. Well-designed internal standards, however, are also an opportunity to accomplish other tasks such as assessing the quality of an assay and enhancing the ability to compare analyses over long periods of time.

To be effective, internal standards must accurately reflect the effects of the sample processing on the recovery of proteins and the ability of the mass spectrometer to measure those proteins in any given experiment. Therefore, it is important that the internal standards be added to the sample as soon as possible, be able to equilibrate with the endogenous proteins in the sample, undergo the same processing steps those proteins undergo, and finally are measured at the same time as the other proteins in the sample. Our laboratory uses a combination of approaches to internal standards—the constant measurement of a small number of endogenous housekeeping proteins and the use of a non-endogenous protein that is added to the sample as the first step after homogenization.

Endogenous housekeeping proteins are proteins found in a sample that do not change under the experimental conditions being studied. The use of housekeeping proteins is a common approach in quantitative analyses in which changes in expression of a protein or set of proteins is demonstrated. The housekeeping proteins are used to show equality in all other aspects of the experiment. In the case of Western blot analysis, the housekeeping protein is sometime called a loading control. A parallel scenario is the use of housekeeping genes in experiments that measure changes messenger RNA, like quantitative PCR. Endogenous housekeeping proteins also meet all the criteria for a good internal standard in a quantitative mass spectrometry experiment. They are an intrinsic part of the proteins in the sample so they have equilibrated with those proteins and must undergo all of the same processing steps. The housekeeping proteins are then measured in the same selected reaction monitoring experiment via an optimized descriptor as described in Chap. 3, just like all of the other proteins in the assay. One crucial advantage of the housekeeping proteins is that because they are already a part of the sample, no additional protein is added. A second major advantage of using housekeeping proteins in a selected reaction monitoring experiment is that other scientists are familiar with and accept the idea that lack of changes in these proteins are a good tool for demonstrating the validity of changes seen in other proteins. In other words, the constant use of housekeeping proteins in Western blot analysis and housekeeping genes in quantitative PCR means that the approach has been tested in uncountable experiments over the years and is accepted by other scientists.

Our laboratory uses two proteins as housekeeping proteins in the β-oxidation assay; HSPD1 and VDAC1. These proteins were initially selected as a part of other on-going experiments investigating the changes in mitochondrial proteins and based on our observations of good abundance and good performance in the selected reaction monitoring experiment. We also rationalized that they *should not* change expression in the types of experiments being done because VDAC1 is a voltage gated anion channel in the outer mitochondrial membrane and HSPD1 is a mitochondrial protein needed for the proper folding of imported into the mitochondria. We recognize that this rationalization has little value beyond the initial design of the assay. The abundance of the proteins was monitored over a number of experiments. Statistical tests were used to demonstrate consistent expression both within experiments and between multiple experiments. As their use became more established in our group, additional experiments were carried out using Western blot to test the lack of change in a number of representative experiments. It is also notable that the BSA internal standard described below gives additional verification that the housekeeping proteins are not changing.

Overall, the choice of housekeeping proteins is highly variable with few guiding principles. Other proteins we have used include calreticulin (CALR), nucleolin (NCL), and 40S ribosomal protein S8 (RPS8) representing a variety in intracellular localizations. We have also used the enzymes glyceraldehyde-3-phosphate dehydrogenase (GAPDH) and lactate dehydrogenase A (LDHA). In all cases, it is important that the constancy of housekeeping proteins be verified. Statistical tests of the mass spectrometry data and complementary quantitative analyses such as Western blot and enzymatic activity assays are appropriate [9]. It should also be recognized that in highly multiplexed assays, the majority of the protein targets will not change [6, 9]. This set of additional unchanged proteins will, in effect, represent a set of *ad hoc* housekeeping proteins.

A non-endogenous standard is a protein that is not found in the samples being analyzed. Our laboratory uses BSA for this purpose in most samples. Care must be taken because albumin is a part of all tissues and BSA is a part of nearly all cell culture media. Our experiments with mouse tissues use BSA and monitor peptides that are only seen in the bovine, as opposed to mouse, form of the protein. In cell culture experiments, we have used chicken egg lysozyme in some experiments but currently prefer equine serum albumin. The use of equine serum albumin in all experiments is a change we expect to make but have not yet done so for the continuity of several on-going experiments.

Advantages of BSA for standard are that it is relatively inexpensive and plentiful. Therefore, making standard solutions with accurately weighed amounts of protein is a straight-forward process. It is also possible to purchase standard solutions of BSA. Albumin produces a number of amenable peptides for the selected reaction monitoring analysis. In any case, the internal standard is added to an aliquot of the sample homogenate immediately after the homogenization steps. Equilibration of the standard with the endogenous proteins is facilitated by denaturing with detergents like SDS and heating. From that point forward, the albumin tracks the endogenous proteins through all subsequent preparation steps. As described above for the housekeeping proteins, the albumin is detected in the selected reaction monitoring assay based on an optimized, multi-peptide descriptor developed using the methods in Chap. 3.

Other laboratories use a variety of stable isotope labeled internal standards. The most common of these standards are the so-called AQUA-peptides. The AQUA term comes from the absolute quantification method for proteins and phosphoproteins described by Gygi and co-workers [10]. These peptides are individual synthesized with an isotope labeled amino acid incorporated. An application of these peptides combined the AQUA-peptides with immunoaffinity isolation in a method called stable isotope standards and capture by anti-peptide antibodies (SISCAPA) [11]. Others have taken the next step of developing stable isotope labeled proteins [12]. Finally there is the QconCAT method that engineers a specific synthetic protein that contains the peptides from multiple proteins and is expressed in conditions that give isotopically label the protein [13]. The important element that stable isotope labeled standards adds is the measurement of both the target peptide and the stable isotope labeled standard peptide at the exact same time in the LC-tandem mass spectrometry experiment because of their co-elution. This simultaneous measurement rigorously compensates for any variation in instrument sensitivity including possible ion suppression due to other closely eluting peptides.

Overall, the use of stable isotope labeled internal standard combined with mass spectrometry is considered the most rigorous type of quantitative assay available. Several potential issues, however, have made their incorporation a low priority for our laboratory. For example, these materials can add a significant expense to the experiment. AQUA-peptides cost approximately $400 *per peptide* for the initial purchase, although the amount of material is likely sufficient for several hundred samples. Similarly, producing isotopically labeled proteins or the QconCAT proteins can be both laborious and expensive. At this time, however, the key issue for our laboratory is the high level of performance that is seen without these standards. The precision of the assay have been sufficient to demonstrate, with appropriate statistical test, changes in the abundance of numerous proteins in the range of 30 % increases or decreases. A general observation would be that with a well-designed non-endogenous internal standard the precision of the selected reaction monitoring experiments is better than the biological variation seen in our animal and cell culture experiments. Therefore, any potential enhancement of precision seen through the use of stable isotope labeled peptides may not translate into a practical enhancement of the overall assay performance.

5.3 Data Analysis

Data analysis strategies follow standard principles of quantitative LC-mass spectrometry experiments, although no specific consensus methods have been described. The central components of the data analysis are (a) determine the mass spectrometric response of the analyte, which in these cases are the peptides acting as markers for the target proteins, (b) determine the mass spectrometric response of the standard

1. Integrate chromatographic peak areas for all peptides

2. Calculate total protein response (TPR) from peptides for each:

   a) target protein
   b) housekeeping protein
   c) internal standard

3. Calculate protein amount

$$\frac{\text{TPR-target or housekeeping}}{\text{TPR-internal standard}} \times \text{amount of internal standard}$$

4. Calculate protein concentration in sample

   protein amount ÷ mg protein in sample

Fig. 5.4 Summary of steps in data analysis and calculations. A simple series of steps are used to convert the integrated chromatographic peak areas of the peptides for each protein into a protein concentration

(or standards), and (c) convert the ratio of the analyte response to the standard response into an amount based on a relative response factor.

The process utilized in our laboratory is summarized in Fig. 5.4. As a start, the program Pinpoint (from ThermoScientific) is used to automatically calculate the integrated peak area of all peptides in the assay. The program uses the alignment of the multiple reactions monitored for each peptide to find the chromatographic peaks. The location of the chromatographic peaks found in a designated reference run are used to direct the recognition of the peptides in all samples in a set, with manual editing of those chromatographic peaks allowed. The peak area integration includes the signal for all fragmentation reactions recorded for each peptide. The integrated peak areas of all peptides recorded for a protein are totaled to give the overall response for that protein. We refer to this value as the total protein response (TPR). As an example from the case of ACOX1 based on the descriptor described in Chap. 3, the integrated peak areas for all three ACOX1 peptides are totaled to give the ACOX1 total protein response. All other proteins are then calculated in a similar manner. At this time, we do not make any specific corrections for the proteins that are detected based on only two peptides. That type of correction is logical and easy, we just have been taking a simpler approach at this time.

The final part of the calculation is to use the ratio of the total protein response for each target protein to the total protein response for the internal standard protein bovine serum albumin and use this ratio to calculate an amount of the target protein in the sample. This amount of protein is the primary result determined by the assay and is based on the best flyer concept discussed during the descriptor design process in Chap. 3 [14]. For example, 8 pmol of bovine serum albumin is added to all samples as the non-endogenous internal standard, so the pmol equivalent of all proteins in the assay is determined. This value is then used to calculate a pmol/mg protein concentration based on the total protein in the sample as measured during the initial sample processing steps. An assumption of equal response for both the target protein and the bovine serum albumin internal standard is made, invoking the best flyer concept.

The strength of this data analysis approach is the inherent simplicity. As with many analytical methods, careful work in the initial design and development process to the best peptides for each protein in the assay creates a foundation that supports this simple approach. The ratio of combined signal for that group of peptides versus the group of peptides for the bovine serum albumin then gives a functional measure of the amount of the target protein in the sample. This measure is ideally suited to any experiment in which changes in the amount of protein in a sample is being compared among different experimental conditions—a *very* common goal in biomedical research. Further, one could argue that it is also likely that this measure also represents an equally good and accurate determination of the absolute amount of a protein in the sample. This argument is likely more contentious and depends on how one views of the assumptions incorporated into the different measurement techniques. The primary assumptions of this best flyer approach are (a) that the best peptides have been identified, (b) those peptides are *average* tryptic peptides with an *average* electrospray response, and (c) the LC-tandem mass spectrometry measurement of those peptides is not adversely affected in any given sample because the target peptides and the standard peptides are not measured at exactly the same time. If so, then the relative response between the target protein and standard protein is 1. Techniques such as the AQUA method must also make assumptions with the most significant that all processing steps are not adversely affected in any given sample, because the isotopically labeled peptide does not model all the processing steps. In either case, absolute quantification of proteins, in the truest sense, is difficult because few isolated and purified proteins are available in forms that could be used to make the requisite primary standards. We would contend that this difficulty in assuring absolute quantification does not hurt the utility of targeted quantitative proteomics in any way.

5.4 Quality Control

A part of all analytical experiments is assessing the quality of the assay. Quality in all phases of the experiment should be measured and monitored from the initial samples, through the sample processing, and into the LC-tandem mass spectrometric analyses.

Assessing the quality of the samples begins with the short run gel electrophoresis described in Chap. 4. The SDS-PAGE gel gives a direct visualization of the sample that allows two components of the sample quality to be monitored by simply inspecting the gel, the equality of the loading and the potential for interfering species. Equal loading is seen by the amount of staining in each lane. No specific tool or algorithm is used to quantify the staining, rather the gel is simply scanned as a part of the overall documentation of the analysis. Large differences are noticeable and would begin some corrective action. Interfering species are most often recognized by streaks in the gel or a pattern of poorly shaped bands. High salt will sometimes cause the lanes to swell. Depending on a number of factors, the corrective actions range from rerunning the gel to correct loading problems to more elaborate efforts to repeat the sample preparation. In some experiments, single problem samples will simply be eliminated from the analysis. The larger point, however, is that care scrutinizing the sample preparation process and the resultant SDS-PAGE gels allows us to recognize problem samples and correct those problems in a way that minimizes time lost to the subsequent sample and data analyses.

Quality is also tracked during the sample analyses though a number of features that are measured in all samples. The most important part of the quality control assessment is the detection of the BSA peptides. Since a consistent amount of the protein, 8 pmol, is added to all samples as an internal standard, a similarly consistent signal is expected for the three peptides being monitored. We have selected three parameters related to the internal standard to monitor and chart for each set of samples, the signal for the peptides LVTDLTK and LVNELTEFAK, and the ratio of those two signals. Finally, the retention time for the VATVSLPR autolysis peptide from trypsin is also tracked. The abundance and ratio of the two BSA peptides are the primary quality control measures. An example of 40 consecutive analyses is shown in Fig. 5.5. These three measures reflect several critical aspects of the assay; how much BSA was added, the complete digestion of the sample, and the appropriate response of the mass spectrometer. The retention time of the VATVSLPR peptide is a primary measure because this retention time is targeted by the flow rate of the LC to adjust the retention characteristics to put the various target peptides in the time windows being monitored for each.

Over time, a series of expectations have been developed for each parameter. The movement of any of the values outside of these expectations has a range of interpretations. The simplest to understand is the need to adjust the flow rate to adjust the retention time observed for the VATVSLPR peptide. The two other common assay maintenance decisions that are made are the status of the capillary LC column and the cleanliness of the ion source. In each case, issues are recognized by combinations of low abundance of the different peptides or significant changes in the abundance ratios. Specific corrective actions are taken based on factors such as the age of the column and time since the last ion source cleaning.

Laboratory automation makes it particularly important to actively monitor quality since the instrument is capable of running large numbers of samples unattended. A simple system of monitoring a few parameters will help minimize lost time and sample amounts when the system is not operating as needed for an analysis. Similarly, the amount of data produced makes data analysis a time-consuming process that should not be wasted on poor quality analyses. Finally, documenting that the instrument system is working correcting can expedite other corrective actions that may be important to the overall success of the experiment, such as resampling animals or reacquiring samples in a timely manner.

Fig. 5.5 Quality control charts monitoring the detection of two peptides from the non-endogenous internal standard protein bovine serum albumin. Abundance of the two peptides (**a**) LVTDLTK and (**b**) LVNELTEFAK are plotted for a series of 40 consecutive quantitative proteomic experiments. (**c**) The ratio of the two peptides is also plotted. The *lines* represent the expectations for each measure based on all samples run over an approximately 2-year period of time

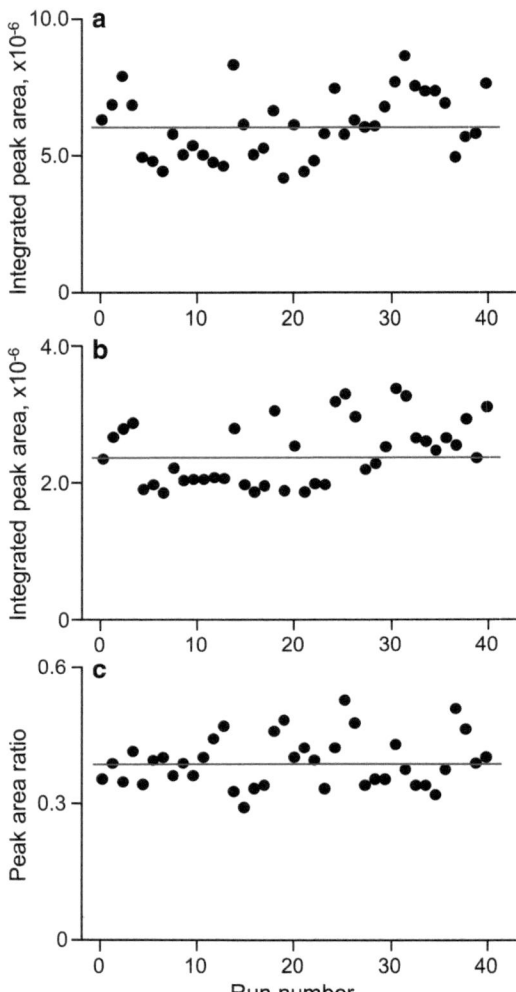

5.5   Summary

The accumulation of multiple peptide descriptors in a multiplexed assay that measures multiple proteins is a critical part of the power of selected reaction monitoring. Like other high sensitivity methods, targeted quantitative proteomics requires some type of internal standard. A couple of simpler options are a non-endogenous internal standard or housekeeping proteins as describe here. More elaborate strategies have been described by others. Building the larger assay also requires attention to ability of the instrument to monitor the requisite number of fragmentation reactions at a rate that effectively reconstructs the chromatographic peak. Finally, it is highly recommended that quality control measures be established that will allow the performance of the assay to be tracked and will help recognize problems before time and sample are lost to poor performing experiments. The ability of more automated systems to run more samples in an unattended manner increases the need for good quality control.

References

1. Anderson L, Hunter CL (2006) Quantitative mass spectrometric multiple reaction monitoring assays for major plasma proteins. Mol Cell Proteomics 5:573–588
2. Keshishian H, Addona T, Burgess M, Mani DR, Shi X, Kuhn E, Sabatine MS, Gerszten RE, Carr SA (2009) Quantification of cardiovascular biomarkers in patient plasma by targeted mass spectrometry and stable isotope dilution. Mol Cell Proteomics 8:2339–2349
3. Kikuchi T, Hassanein M, Amann JM, Liu Q, Slebos RJ, Rahman SM, Kaufman JM, Zhang X, Hoeksema MD, Harris BK, Li M, Shyr Y, Gonzalez AL, Zimmerman LJ, Liebler DC, Massion PP, Carbone DP (2012) In-depth proteomic analysis of nonsmall cell lung cancer to discover molecular targets and candidate biomarkers. Mol Cell Proteomics 11:916–932
4. Barnidge DR, Goodmanson MK, Klee GG, Muddiman DC (2004) Absolute quantification of the model biomarker prostate-specific antigen in serum by LC-MS/MS using protein cleavage and isotope dilution mass spectrometry. J Proteome Res 3:644–652
5. Williams DK, Muddiman DC (2009) Absolute quantification of C-reactive protein in human plasma derived from patients with epithelial ovarian cancer utilizing protein cleavage isotope dilution mass spectrometry. J Proteome Res 8:1085–1090
6. Rindler PM, Plafker SM, Szweda LI, Kinter M (2013) High dietary fat selectively increases catalase expression within cardiac mitochondria. J Biol Chem 288:1979–1990
7. Crewe C, Kinter M, Szweda LI (2013) Immediate inhibition of pyruvate dehydrogenase: An initiating event in high dietary fat-induced loss of metabolic flexibility in the heart (Submitted)
8. Shuford CM, Li Q, Sun YH, Chen HC, Wang J, Shi R, Sederoff RR, Chiang VL, Muddiman DC (2012) Comprehensive quantification of monolignol-pathway enzymes in Populus trichocarpa by protein cleavage isotope dilution mass spectrometry. J Proteome Res 11:3390–3404
9. Kinter CS, Lundie JM, Patel H, Rindler PM, Szweda LI, Kinter M (2012) A quantitative proteomic profile of the Nrf2-mediated antioxidant response of macrophages to oxidized LDL determined by multiplexed selected reaction monitoring. PLoS One 7:e50016
10. Gerber SA, Rush J, Stemman O, Kirschner MW, Gygi SP (2003) Absolute quantification of proteins and phosphoproteins from cell lysates by tandem MS. Proc Natl Acad Sci USA 100:6940–6945
11. Anderson NL, Anderson NG, Haines LR, Hardie DB, Olafson RW, Pearson TW (2004) Mass spectrometric quantitation of peptides and proteins using Stable Isotope Standards and Capture by Anti-Peptide Antibodies (SISCAPA). J Proteome Res 3:235–244
12. Hanke S, Besir H, Oesterhelt D, Mann M (2008) Absolute SILAC for accurate quantitation of proteins in complex mixtures down to the attomole level. J Proteome Res 7:1118–1130
13. Rivers J, Simpson DM, Robertson DH, Gaskell SJ, Beynon RJ (2007) Absolute multiplexed quantitative analysis of protein expression during muscle development using QconCAT. Mol Cell Proteomics 6:1416–1427
14. Ludwig C, Claassen M, Schmidt A, Aebersold R (2012) Estimation of absolute protein quantities of unlabeled samples by selected reaction monitoring mass spectrometry. Mol Cell Proteomics 11:M111.013987

Chapter 6
Future Directions

Abstract The future directions for the targeted quantitative proteomics experiment should address a small number of current problems and incorporate new developments in mass spectrometry instrumentation. Two notable problems at this time are the broader acceptance of selected reaction monitoring as a primary method of protein quantitation and at least some improvement in the sensitivity of the analysis. Improvements will always include the continuing evolution of the instruments to have higher resolution, faster scan rates, and better sensitivity. Others improvements will be seen in continued development of the peptide mass spectrometry databases. Progress in each of these areas is inevitable. The most difficult area to predict is the potential for new chemical tools. For example, reagents that improve sample processing or increase the sensitivity of selected classes of peptides can also have a major impact.

6.1 Wider Acceptance of Selected Reaction Monitoring for Protein Quantitation

As a mass spectroscopist, one of the most difficult to understand issues facing targeted quantitative proteomics using LC-tandem mass spectrometry is the limited acceptance of the method by the scientific community. For example, a common question for proteomics results has always been the question of *validation*, often with the suggestion that a Western blot be used as some type of highly accurate standard method to test an observation made with mass spectrometry. While Western blot is indeed a powerful technique with an unquestionable role biomedical investigation, all users are aware that few available antibodies have been validated beyond a simple test of recognizing a protein at an approximately correct molecular weight position. Unfortunately, all Western blot users also have firsthand experience with antibodies that do not work in their hands, recognize multiple bands, or do not recognize the correct band. Mass spectroscopist can argue that the selected reaction

M. Kinter and C.S. Kinter, *Application of Selected Reaction Monitoring to Highly Multiplexed Targeted Quantitative Proteomics*, SpringerBriefs in Systems Biology, DOI 10.1007/978-1-4614-8666-4_6, © The Authors 2013

monitoring experiment, through the steps needed to develop a descriptor, endure more rigorous testing of each protein analysis than the antibodies used for Western blot. The inherent amino acid sequence information, the m/z of the parent ion and product ions being specified to a tenth of a Dalton or better, precise chromatographic retention times that can be predicted based on the amino acid content, and the specificity of the trypsin all represent peptide specific and, therefore, protein specific parameters. As shown in Fig. 3.6, it is also possible to compare reconstructed CID spectra to library spectra as an additional test of detecting the correct peptide. For the most demanding assay, co-chromatography with synthetic stable isotope labeled standards is also available. With this in mind, one would argue that the Western blot analysis is best seen as a potential secondary test that is consistent with careful science, but should not be considered a primary standard against which a mass spectrometry result is judged.

The larger question of the acceptance of selected reaction monitoring as an important tool for protein quantitation is ability to move the method out of the research laboratory into regulated areas such as the clinical laboratory. A number of published results showing the ability to measure clinically relevant proteins by selected reaction monitoring are just the first step in this process [1–3]. The growing recognition that LC-tandem mass spectrometry should be the next step in the validation of new biomarkers should continue this process [4]. We have argued that the implementation step that comes after validation should continue to use LC-tandem mass spectrometry [5]. In this regard, the National Cancer Institute has established the Office for Cancer Clinical Proteomics Research (http://proteomics.cancer.gov/). Parts of this group have been evaluating the issues faced in using mass spectrometric methods for new molecular diagnostics [6]. The efforts of this group have lead to the submission of a mock submission to the Food and Drug Administration [7]. A bottom line in all of these endeavors is that multiplexed targeted quantitative proteomics remain a relatively new analytical method. One part of this *newness* is that we do not yet have a full understanding of all the issues. Results and experience will come quickly to both identify and solve these issues to the standards needed for new molecular diagnostic systems.

6.2 Databases that Share Information Will Expand the Application

Resources like the PeptideAtlas give one instant access to the cumulative efforts of large numbers of scientists. This effort greatly facilitates the development process as described in Chap. 3. Further, the availability of these libraries of CID spectra can also function as an important proof for the proper detection of the desired peptide. The efforts in database development and growth extend to the final selected reaction monitoring methods. The Institute for Systems Biology also maintain the SRMAtlas database (http://www.srmatlas.org/). As of June 2013, this database is extensive for yeast but is awaiting the release of similar human and mouse data [8]. A similar glycopeptide database has recently been announced [9]. As experiments continue to show the transferability of methods, the value of this type of database will become even more apparent.

6.3   What Will Advances in Instrumentation Add?

In the field of mass spectrometry, new and improved instrumentation is a constant story. Driven by both the scientist that use mass spectrometry and the scientists and engineers working for the instrument manufacturers, the evolution of the instrumentation has played a dominant role in keeping the field at the forefront of analytical chemistry. Three crucial areas that have seen continuous improvement are sensitivity, resolution, and speed of the acquisition. Improvements in sensitivity and resolution, in particular, will directly benefit targeted quantitative proteomics.

Advancing sensitivity will likely have the most profound effect. A common misconception is that mass spectrometry is more sensitive than immunochemical methods like Western blot. The fact is that, unless the antibody is performing far below average, Western blot analysis is more sensitive that mass spectrometry. The sensitivity of immunochemical methods is particularly apparent in the detection of phosphorylated proteins, especially considering the low stoichiometry of most phosphorylation events. In these cases, detection of the phosphorylated protein by Western blot with a properly performing antibody is routine. The demonstration of changes in phosphorylation by simultaneously showing no change in unmodified parent protein can be clear and powerful.

The gap between mass spectrometry and immunochemical methods, however, is continuously closing with each new generation of mass spectrometry instrumentation.

Higher resolution, particularly in more moderately priced instrument systems, will add to targeted quantitative proteomics in at least two ways. First, higher resolution for both the parent m/z and fragment ion m/z is an additional enhancement of specificity of the analysis. Second, it is possible that higher resolution mass spectrometry system will allow broader evaluation of selected ion monitoring (SIM) for targeted quantitative proteomics. The new instruments also introduce new scan modes for the experiment. For example, the geometry of some on these new high resolution instruments, specifically those instruments with a C-trap for ion accumulation and an orbitrap for the high resolution mass analysis, have the ability to record complete high resolution CID spectra for each precursor ion. These experiments have been called parallel reaction monitoring because ions from one precursor ion are being selected and fragmented while a preceding group of fragment ions are being mass analyzed [10].

6.4   Chemical Approaches and Reagents Are an Opportunity

A final area with unknown potential is the development of novel chemical approaches and reagents. Proteomics already has a strong history in this regard. Tools likes isotope coded affinity tags and stable isotope labeling in culture are prime examples. Not only has targeted quantitative proteomics made good use of these tools, but the application has its own examples. The isolation of specific targets, as done with SISCAPA, is clearly an area that can directly influence the sensitivity of the technique for specific targets. The continued development of aptamers and other novel isolation reagents would be welcome. Refined internal standard methods like QconCAT add to the accuracy of the method. Finally, there would also appear to be a need for reagents that enhance the sensitivity of specific classes of peptides. One such reagent is the ALiPHAT reagent (augmented limits of detection for peptides with hydrophobic alkyl tags) used to label cysteine containing peptides [11]. The development of other enhancement reagents for phosphopeptides, as an example, would have many benefits. Overall, while this area is harder to define, it might ultimately represent an area with unique opportunities.

6.5   Summary

The recent recognition of targeted quantitative proteomics as the Method of the Year for 2012 by the journal Nature Methods is a significant recognition. Our experience, and that of colleagues also using the method, is that selected reaction monitoring for targeted quantitative proteomics works very well. The methods and the instrument system are robust, reliable, and very productive. Wider application will naturally illuminate currently unseen problems and drive new solutions to those problems.

References

1. Barnidge DR, Goodmanson MK, Klee GG, Muddiman DC (2004) Absolute quantification of the model biomarker prostate-specific antigen in serum by LC-MS/MS using protein cleavage and isotope dilution mass spectrometry. J Proteome Res 3:644–652
2. Kuhn E, Addona T, Keshishian H, Burgess M, Mani DR, Lee RT, Sabatine MS, Gerszten RE, Carr SA (2009) Developing multiplexed assays for troponin I and interleukin-33 in plasma by peptide immunoaffinity enrichment and targeted mass spectrometry. Clin Chem 55:1108–1117
3. Hoofnagle AN, Becker JO, Wener MH, Heinecke JW (2008) Quantification of thyroglobulin, a low-abundance serum protein, by immunoaffinity peptide enrichment and tandem mass spectrometry. Clin Chem 54:1796–1804
4. Whiteaker JR, Lin C, Kennedy J, Hou L, Trute M, Sokal I, Yan P, Schoenherr RM, Zhao L, Voytovich UJ, Kelly-Spratt KS, Krasnoselsky A, Gafken PR, Hogan JM, Jones LA, Wang P, Amon L, Chodosh LA, Nelson PS, McIntosh MW, Kemp CJ, Paulovich AG (2011) A targeted

proteomics-based pipeline for verification of biomarkers in plasma. Nat Biotechnol 29:625–634

5. Kinter M (2004) Toward broader inclusion of liquid chromatography-mass spectrometry in the clinical laboratory. Clin Chem 50:1500–1502

6. Rodriguez H, Tezak Z, Mesri M, Carr SA, Liebler DC, Fisher SJ, Tempst P, Hiltke T, Kessler LG, Kinsinger CR, Philip R, Ransohoff DF, Skates SJ, Regnier FE, Anderson NL, Mansfield E, Workshop Participants (2010) Analytical validation of protein-based multiplex assays: a workshop report by the NCI-FDA interagency oncology task force on molecular diagnostics. Clin Chem 56:237–243

7. Regnier FE, Skates SJ, Mesri M, Rodriguez H, Tezak Z, Kondratovich MV, Alterman MA, Levin JD, Roscoe D, Reilly E, Callaghan J, Kelm K, Brown D, Philip R, Carr SA, Liebler DC, Fisher SJ, Tempst P, Hiltke T, Kessler LG, Kinsinger CR, Ransohoff DF, Mansfield E, Anderson NL (2010) Protein-based multiplex assays: mock presubmissions to the US Food and Drug Administration. Clin Chem 56:165–171

8. Picotti P, Lam H, Campbell D, Deutsch EW, Mirzaei H, Ranish J, Domon B, Aebersold R (2008) A database of mass spectrometric assays for the yeast proteome. Nat Methods 5:913–914

9. Hüttenhain R, Surinova S, Ossola R, Sun Z, Campbell D, Cerciello F, Schiess R, Bausch-Fluck D, Rosenberger G, Chen J, Rinner O, Kusebauch U, Hajdúch M, Moritz RL, Wollscheid B, Aebersold R (2013) N-glycoprotein SRMAtlas: a resource of mass spectrometric assays for N-glycosites enabling consistent and multiplexed protein quantification for clinical applications. Mol Cell Proteomics 12:1005–1016

10. Peterson AC, Russell JD, Bailey DJ, Westphall MS, Coon JJ (2012) Parallel reaction monitoring for high resolution and high mass accuracy quantitative, targeted proteomics. Mol Cell Proteomics 11:1475–1488

11. Frahm JL, Bori ID, Comins DL, Hawkridge AM, Muddiman DC (2007) Achieving augmented limits of detection for peptides with hydrophobic alkyl tags. Anal Chem 79:3989–3995